EINSTEIN

A HUNDRED YEARS OF RELATIVITY

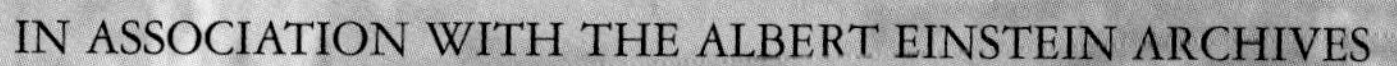
IN ASSOCIATION WITH THE ALBERT EINSTEIN ARCHIVES

EINSTEIN

A HUNDRED YEARS OF RELATIVITY

ANDREW ROBINSON

With contributions by:
Philip Anderson, Arthur C. Clarke, I. Bernard Cohen,
Freeman Dyson, Philip Glass, Stephen Hawking,
Max Jammer, João Magueijo, Joseph Rotblat,
Robert Schulmann, Steven Weinberg

Harry N. Abrams, Inc., Publishers

In memory of my father, F. N. H. Robinson,
experimental and theoretical physicist,
who, like Einstein, loved music and sailing

Produced by Palazzo Editions Limited, 15 Gay Street, Bath BA1 2PH, UK
In association with the Albert Einstein Archives at the Hebrew University of Jerusalem
Publisher: Oliver Craske
Designed by Nadine Levy and David Costa at Wherefore Art?, London

Library of Congress Cataloging-in-Publication Data

Robinson, Andrew.
Einstein : a hundred years of relativity / by Andrew Robinson ; with contributions from Freeman Dyson ... [et al.].
p. cm.
Includes bibliographical references and index.
ISBN 0-8109-5923-2 (hardcover : alk. paper) 1. Einstein, Albert, 1879-1955. 2. Physicists—Biography.
3. General relativity (Physics) 4. Physics—History—20th century. I. Title.

QC16.E5R63 2005
530'.092—dc22

2005006593

Published in 2005 by Harry N. Abrams, Incorporated, New York

Printed in Singapore
10 9 8 7 6 5 4 3 2 1

Harry N. Abrams, Inc.
100 Fifth Avenue
New York, N.Y. 10011
www.abramsbooks.com

Abrams is a subsidiary of
LA MARTINIÈRE
GROUPE

Pages 2–3: Albert Einstein on top of the Empire State Building, New York, late 1930s.

Opposite: Einstein in 1946 in Washington, DC at a hearing of the Anglo-American Committee, which was investigating disturbances between Arabs and Jews in the Holy Land.

Page 7: Einstein walking through the campus of Princeton University, New Jersey, 1953.

Contents

Part One: The Physicist

Part Two: The Man

Preface

Freeman Dyson

Albert Einstein's life was full of paradoxes. This book, published in the centenary of his 'miraculous year,' 1905, describes his life and work and personality, all of which were paradoxical in one way or another. Articles written by experts discuss Einstein's views on space and time, chance and necessity, religion and philosophy, marriage and politics, war and peace, fame and fortune, life and death. I am not an expert on Einstein, but I know something about the universe, so I discuss his view of the universe. Einstein's universe was paradoxically different from ours. It had no black holes.

Black holes—first mooted as far back as 1783 by an English astronomer—are familiar objects to today's astronomers. We know that they exist all over our own galaxy and in the central regions of other galaxies. We see them as sources of X-ray radiation emitted by gas as it falls into them and is heated to temperatures of millions of degrees by their overwhelmingly strong gravity. At the exact centre of our own galaxy we see a black hole weighing as much as a few million suns, with massive stars orbiting around it like moths around the flame of a candle. From time to time, roughly once in 10,000 years, one of the moths will fly into the flame and burn. One of the orbiting stars will pass too close to the black hole and will be torn into an orbiting tangle of spaghetti by tidal forces stronger than its own feeble gravity. The orbiting star will die, part of it being swallowed by the black hole and the rest blown away as an expanding cloud of gas and X-rays. Black holes are not rare, and they are not an accidental embellishment of our universe. They are a fundamental driving force of its evolution. They are the dominant source of energy. For every ounce of matter consumed, black holes yield about a hundred times more energy than the nuclear reactions that cause our Sun to shine and our hydrogen bombs to explode. To a modern astronomer, a universe without black holes makes no sense.

To a modern physicist, black holes are also objects of transcendent beauty. They are the only places in the universe where Einstein's theory of general relativity shows its full power and glory.

Here, and nowhere else, space and time lose their individuality and merge together into a sharply curved four-dimensional structure precisely delineated by Einstein's equations. If you were to imagine yourself falling into a black hole, your local perception of space and time would be detached from the space and time of an observer watching you from outside. While you would see yourself falling smoothly into the hole without any deceleration, the outside observer would see you coming to a halt at the horizon of the hole and remaining for ever in a state of permanent free fall. Permanent free fall is a situation that can only exist by virtue of the distortion of space and time predicted by Einstein's theory. As seen from the outside, you would keep on falling for ever into the hole and never reach the bottom.

Black holes can also rotate, and the behaviour of space and time as you fall into a rotating black hole is even more peculiar. Rapidly rotating black holes can be a prodigious source of energy. The gamma-ray bursts which we see at a rate of about one per day in remote parts of the universe are the most violent of all natural events; and the most plausible theory to explain them has them originating as instabilities of rotating black holes. All these strange and beautiful features of our universe would have been unimaginable without Einstein's theory to guide our thinking.

And still the paradox remains. Einstein repudiated black holes and declared them, in a famous paper published in 1939, non-existent. It appeared in the leading American mathematics journal, *Annals of Mathematics*, and attracted much attention. Einstein constructed a very artificial model of a static black hole, with a cloud of particles of matter orbiting in a hollow spherical shell held together by their mutual gravitational attraction. He showed that this model was impossible because the particles on the outside of the shell would have to travel faster than light. He then concluded by saying, "The essential result of this investigation is a clear understanding of why the Schwarzschild singularities do not exist in physical reality." "Schwarzschild singularity" was the phrase used in those days for the object which was later to be called a black hole. Einstein concluded from the failure of this one unrealistic model that no consistent model of a black hole could exist.

His conclusion is a logical non-sequitur. Somehow, Einstein had acquired a deep emotional aversion to the idea of black holes and used this illogical argument to buttress his intuition that they ought not to exist. We now know that his argument was irrelevant because real black holes are not static. They are made by the gravitational collapse of massive objects. They are dynamic objects in a state of permanent free fall.

Einstein never changed his mind on this. He not only believed that the theory of black holes was wrong, he was not even interested in examining the evidence to see whether they might exist in nature—in contrast to his attitude to his celebrated prediction of the bending of light by the Sun, confirmed by the solar eclipse measurements of 1919. His lack of interest is particularly puzzling, because in the same year, 1939, in which he published his denunciation of black holes, J. Robert Oppenheimer and Hartland Snyder published a paper describing in detail how a massive star that had exhausted its nuclear fuel would naturally collapse into a black hole as a consequence of Einstein's equations. Einstein must have known about the Oppenheimer-Snyder calculation, but he never responded to it. A few years later, when Oppenheimer came to Princeton as director of the Institute for Advanced Study, he met Einstein frequently and had plenty of opportunity to talk to him about black holes. So far as I know, the subject was never mentioned.

We now know that the Oppenheimer-Snyder calculation is substantially correct and describes what actually happens to massive stars at the end of their lives. It explains why black holes are abundant, and incidentally confirms the truth of general relativity. And still, Einstein was not interested. The question remains: how could he have been blind to one of the greatest triumphs of his own theory? I have no answer to this question. It remains one of the inexplicable paradoxes in the life of a genius.

Einstein by Low, 1929.

MEMOIRS BY VISCOUNT

1. The World of Physics Before Einstein

"In one person [Newton] combined the experimenter, the theorist, the mechanic and, not least, the artist in exposition."

Einstein in his foreword to Sir Isaac Newton's *Opticks*, 1931

"The whole of science" Albert Einstein wrote in 1936, when he was already regarded as the greatest scientist of his time, "is nothing more than a refinement of everyday thinking." It was a mischievous paradox typical of a man with a genius for discovering simplicity in complexity. True if you were Einstein, maybe—but for most of us it is a scarcely credible statement. Come off it, one cannot help thinking. What have our everyday thought processes got to do with the thinking of great scientists, especially the esoteric mathematical subtleties of twentieth-century physicists?

Physics, in its long evolution towards unifying more and more of the universe on the basis of fewer and fewer fundamental ideas, seems to have moved further and further away from everyday thinking with each passing decade. Most non-physicists get to use some of the technological by-products of pure research in physics: computers, DVD players, mobile phones and the like. But general relativity, the theory that explains black holes (and the accuracy of the satellite GPS), and the quantum theory that underlies superstrings (and

The title page from Galileo Galilei's *Dialogue Concerning the Two Chief World Systems* (1632) depicts (left to right) Aristotle, Ptolemy and Copernicus discussing astronomical matters. Ptolemy holds an armillary sphere in his right hand, while Copernicus holds a representation of the new heliocentric system. An arrow, barely visible on the ground to the left of the publisher's seal, points to Copernicus.

lasers)—both of which have Einstein as a founding father—appear to have nothing in common with everyday experience. Earlier scientific ideas, such as Archimedes' principle, Isaac Newton's laws of motion and gravitation and Michael Faraday's concept of the magnetic field, are relatively accessible to everyday thinking; we can even do simple experiments at home to demonstrate the truth of them with objects immersed in water, falling coins and moving compass needles. Not so with relativity and quantum theory.

Modern science, of course, owes much to the ancient Greeks such as Archimedes, Euclid and Democritus. To the Greek mathematicians and natural philosophers who thought for themselves two millennia ago we owe, for example, the invention of geometry, the idea that light travels in straight lines, the first estimate of the Earth's circumference and the perception that matter is made of atoms. Here the ancients' thinking was marvellously fruitful.

However they also (with one or two enlightened exceptions such as Aristarchus) believed that the planets revolved around the Earth in perfect circles, and that the more massive a body was, the faster it would fall when dropped. Aristotle's 'everyday thinking,' it would seem, led him to the conclusion in his *Mechanics* that "The moving body comes to a standstill when the force which pushes it along can no longer so act to push it"—a grossly inaccurate conception of mass and force. A more massive body should fall faster because, said Aristotle, it had a greater tendency to seek the centre of the Earth—which is also simple to demonstrate as wrong. His concept of 'motion' included not only pushing and pulling but also combining and separating and waxing and waning. A fish swimming and an apple falling from a tree were, obviously enough, in motion—but so too were a child growing into an adult and a fruit ripening. Thus common sense led Aristotle, who was no experimentalist (unlike Archimedes), into a hopeless conceptual muddle about the simplest facts of mechanics.

Yet such was the prestige of the Greeks in philosophy that Aristotle's ideas about the physical world dominated European intellectual life right up to the time of Newton in the seventeenth century, and even beyond

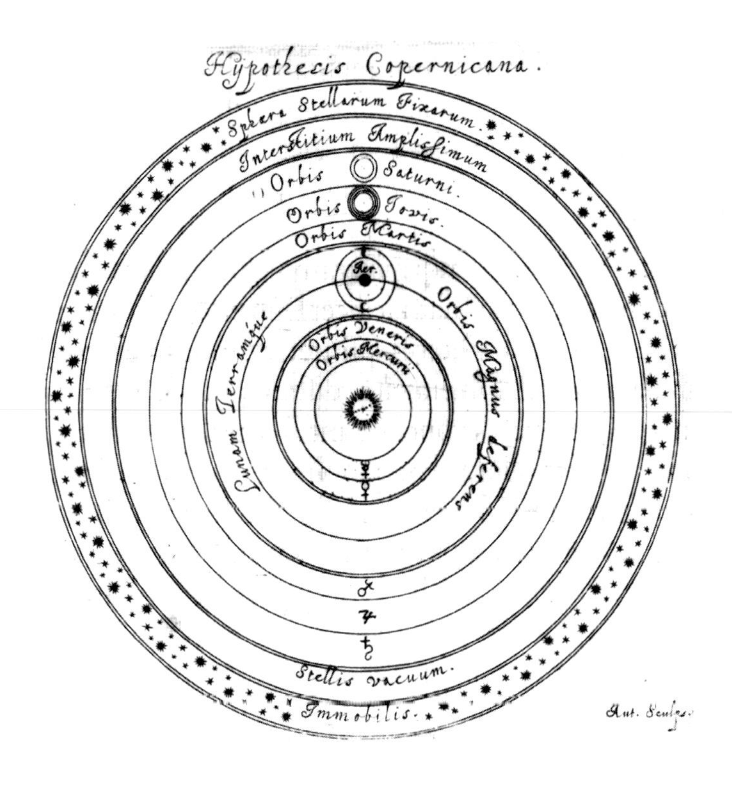

Copernicus's heliocentric cosmos, as depicted in Johannes Hevelius's *Selenographia* (1647). Note that the orbits of the planets are shown as circles rather than as the ellipses established by Johannes Kepler.

in some circles of academia. A few decades before Newton's birth in 1642, the natural philosopher Francis Bacon, whose writings were soon to be instrumental in the launch of the Royal Society, wrote critically that, "All the philosophy of nature which is now received, is either the philosophy of the Grecians, or that other of the alchemists... The one is gathered out of a few vulgar observations, and the other out of a few experiments of a furnace. The one never faileth to multiply words, and the other ever faileth to multiply gold."

But challenges to Aristotle's view of the world were under way. In 1543, on his deathbed, Nicolaus Copernicus published *De Revolutionibus*, his heliocentric picture of the solar system with the Earth and other planets revolving around the Sun. Despite some strong resistance from the orthodox, the idea caught on that the Earth was no longer at the centre of the world (which "taught man to be modest," said Einstein), as it had been since the time of the Greeks, although Copernicus still adhered to the ancient view that the planetary orbits were circular.

Then in 1609 Johannes Kepler, using the first accurate observational data on the planetary motions,

Tycho Brahe, from the title page of his *Astronomiae Instauratae Mechanica* (1598).

Right: Kepler, painted by Hans von Aachen before 1612.

compiled by Tycho Brahe, made an educated guess that their orbits were not circles but ellipses—one of the geometrical forms discovered by the Greeks—and with this mental leap he conceived his laws of planetary motion. These enabled Kepler to calculate astronomical tables, and hence the positions of the planets at any time in the past, present or future, which fitted well with astronomers' observations. As Einstein remarked in 1930, on the tercentenary of Kepler's death, "It seems that the human mind has first to construct forms independently before we can find them in things. Kepler's marvellous achievement is a particularly fine example of the truth that knowledge cannot spring from experience alone but only from the comparison of the inventions of the intellect with observed fact."

At around the same time as Kepler, Galileo Galilei overturned Aristotle's erroneous notions of motion through physical, quantitative experiments with moving objects and falling weights. Galileo showed that a body moving at a constant speed—i.e. uniformly—does not require to be 'pushed' by a force, as Aristotle claimed. For example, a marble set rolling at a given speed on a perfectly horizontal frictionless floor will continue to move at that speed. (In the real world the force of friction will eventually bring it to stop.) And he demonstrated that the speed of a freely falling body does not depend on its mass. Hence the cannon balls of different mass that Galileo allegedly dropped together from the Leaning Tower of Pisa were found to hit the ground at the same time, not at different times (as Aristotle would have predicted). "Pure logical thinking cannot yield us any knowledge of the empirical world; all knowledge of reality starts from experience and ends in it. Propositions arrived at by purely logical means are completely empty as regards reality," wrote Einstein some three centuries later. "Because Galileo saw this, and particularly because he drummed it into the scien-

tific world, he is the father of modern physics—indeed, of modern science altogether."

Einstein had particular reason to admire the Italian physicist, because it was Galileo who was the very first scientist to state the validity of the mechanical principle of relativity. Galileo puts it so beautifully in a 'thought' experiment conducted in his *Dialogue Concerning the Two Chief World Systems* of 1632, that his description is worth quoting in full:

Shut yourself up with some friend in the main cabin below decks on some large ship, and have with you there some flies, butterflies, and other small flying animals. Have a large bowl of water with some fish in it; hang up a bottle that empties drop by drop into a narrow-mouthed vessel beneath it. With the ship standing still, observe carefully how the little animals fly with equal speed to all sides of the cabin. The fish swim indifferently in all directions; the drops fall into the vessel beneath; and, in throwing something to your friend, you need throw it no more strongly in one direction than another, the distances being equal; jumping with your feet together, you pass equal spaces in every direction. When you have observed all these things carefully (though there is no doubt that when the ship is standing still everything must happen in this way), have the ship proceed with any speed you like, so long as the motion is uniform and not fluctuating this way or that. You will discover not the least change in all the effects named, nor could you tell from any of them whether the ship was moving or standing still.

In other words, the unmoving passengers on the moving ship have a velocity relative to the land, but relative to the ship they have no velocity. With reference to the ship they are at rest and feel no force acting upon them, provided that the ship is not accelerating forwards (or turning) and provided they stay below decks away from air currents. For a modern equivalent, think of long-distance air travel in a jet at a speed of many hundreds of miles per hour. During most of the flight at high altitude, unless there is bad weather, while seated and not looking out of the window one has hardly any physical sensation of the aircraft's movement, and walking up the aisle feels the same as walking down

64 *Tabularum Rudolphi*

Tabula Æquationum MARTIS.

Anomalia Eccentri, Cum æquatione parte physica	Intercolumnium, Cum Logarithmo	Anomalia coæquata	Intervallū Cum Logarithmo		Anomalia Eccentri, Cum æquatione parte physica	Intercolumnium, Cum Logarithmo	Anomalia coæquata	Intervallū Cum Logarithmo	
120 4.55.50	8890 1. 5.36	115.17.11	145293 37357		150 2.39.14	16230 1.10.35	147.13.44	140127 33738	
121 4.33. 1	9190 1. 5.47	116.19.52	145080 37211	25	151 2.34.23	16410 1.10.43	148.18.42	140005 33651	15
122 4.30. 6	9480 1. 5.59	117.22.39	144871 37067	25	152 2.29.29	16580 1.10.49	149.23.44	139887 33566	15
123 4.27. 6	9780 1. 6.11	118.25.31	174663 36924	25	153 2.24.33	16750 1.10.56	150.28.49	139773 33484	14
124 4.24. 2	10070 1. 6.22	119.28.29	144458 36782	25	154 2.19.34	16910 1.11. 3	151.33.57	139663 33406	14
125 4.20.53	10360 1. 6.34	120.21.33	144255 36642	25	155 2.14.33	17060 1.11.10	152.39. 9	139558 33331	14
126 4.17.40	10650 1. 6.46	121.34.42	144055 36503	24	156 2. 9.30	17210 1.11.16	153.44.23	139456 33255	13
127 4.14.22	10930 1. 6.57	122.37.56	143857 36365	24	157 2. 4.25	17350 1.11.22	154.49.40	139358 33167	13
128 4.10.59	11210 1. 7. 8	123.41.14	143661 36229	24	158 1.59.18	17480 1.11.28	155.55. 0	139263 33119	12
129 4. 7.31	11480 1. 7.19	124.44.37	143468 36095	24	159 1.54. 8	17600 1.11.33	157. 0.23	139173 33054	12
130 4. 3.58	11740 1. 7.30	125.48. 6	143278 35962	24	160 1.48.56	17720 1.11.38	158. 5.49	139087 32992	11
131 4. 0.21	12000 1. 7.40	126.51.40	143091 35831	24	161 1.43.42	17840 1.11.43	159.11.17	139005 32933	11
132 3.56.40	12260 1. 7.50	127.55.19	142906 35702	23	162 1.38.26	17950 1.11.48	160.16.47	138927 32877	10
133 3.52.55	12510 1. 8. 1	128.59. 3	142724 35575	23	163 1.33. 8	18060 1.11.53	161.22.19	138852 32824	10
134 3.49. 6	12760 1. 8.11	130. 2.52	142545 35450	23	164 1.27.48	18160 1.11.57	162.27.53	138782 32773	9
135 3.45.13	13000 1. 8.21	131. 6.45	142370 35327	23	165 1.22.27	18260 1.12. 2	163.33.29	138716 32725	9
136 3.41.16	13240 1. 8.30	132.10.43	142198 35206	22	166 1.17. 4	18350 1.12. 6	164.39. 6	138654 32681	8
137 3.37.15	13480 1. 8.40	133.14.46	142028 35086	22	167 1.11.46	18440 1.12.10	165.44.45	138597 32640	8
138 3.33.10	13710 1. 8.49	134.18.53	141861 34968	21	168 1. 6.15	18520 1.12.13	166.50.26	138544 32602	7
139 3.29. 1	13940 1. 8.59	135.23. 4	141697 34852	21	169 1. 0.48	18590 1.12.16	167.56. 8	138495 32566	7
140 3.24.43	14160 1. 9. 8	136.27.20	141537 34739	21	170 0.55.20	18650 1.12.19	169. 1.52	138450 32533	6
141 3.20.31	14380 1. 9.16	137.31.41	141381 34628	20	171 0.49.51	18710 1.12.21	170. 7.37	138410 32504	6
142 3.16.10	14600 1. 9.27	138.36. 6	141228 34520	20	172 0.44.21	18770 1.12.24	171.13.24	138374 32478	5
143 3.11.46	14820 1. 9.35	139.40.34	141078 34414	19	173 0.38.50	18820 1.12.26	172.19.12	138341 32455	4
144 3. 7.18	15030 1. 9.44	140.45. 7	140931 34310	19	174 0.33.18	18870 1.12.28	173.25. 0	138313 32434	4
145 3. 2.46	15240 1. 9.54	141.49.44	140788 34209	18	175 0.27.16	18910 1.12.29	174.30.49	138289 32417	5
146 2.58.10	15450 1.10. 2	142.54.24	140649 34110	18	176 0.22.43	18940 1.12.31	175.36.39	138269 32403	3
147 2.53.31	15650 1.10.10	143.59. 8	140513 34013	17	177 0.16.40	18960 1.12.32	176.42.29	138254 32392	2
148 2.48.45	15850 1.10.19	145. 3.56	140381 33919	17	178 0.11. 7	18980 1.12.33	177.48.19	138244 32385	1
149 2.44. 2	16040 1.10.27	146. 8.48	140252 33827	16	179 0. 5.34	18990 1.12.34	178.54.10	138237 32380	1
150 2.39.14	16230 1.10.35	147.13.44	140127 33738	16	180 0. 0. 0	18990 1.12.34	180. 0. 0	138234 32379	0

Tab.Lat.

A page from Kepler's Rudolphine Tables, first published in 1627 in Ulm—the birthplace of Einstein some 250 years later. Based on Tycho Brahe's observations, Kepler's system, with its elliptical planetary orbits around the Sun, allowed planetary positions to be accurately calculated for any given time in the past or future. This page refers to Mars.

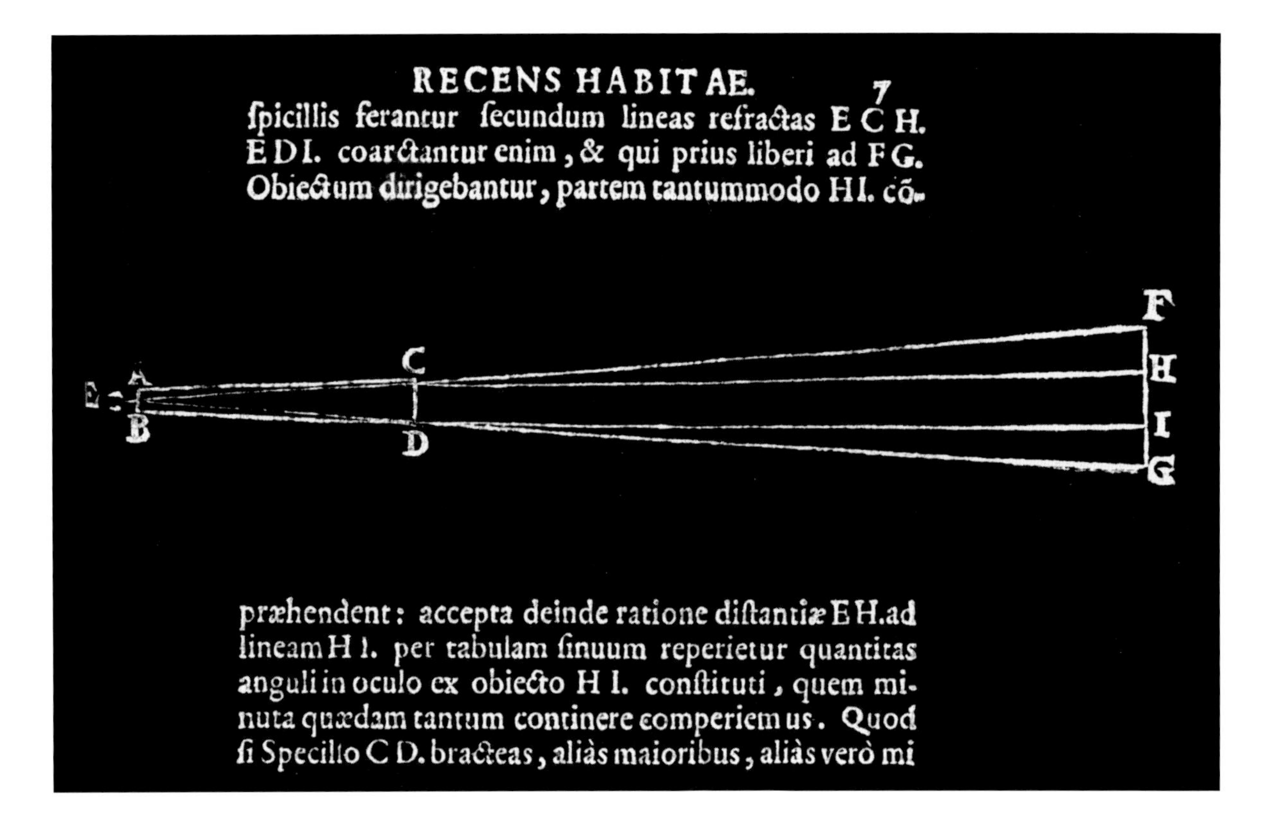

RECENS HABITAE. 7

ſpicillis ferantur ſecundum lineas refractas E C H.
E D I. coarctantur enim, & qui prius liberi ad F G.
Obiectum dirigebantur, partem tantummodo H I. cō-

præhendent: accepta deinde ratione diſtantiæ E H. ad
lineam H I. per tabulam ſinuum reperietur quantitas
anguli in oculo ex obiecto H I. conſtituti, quem mi-
nuta quædam tantum continere comperiemus. Quod
ſi Specillo C D. bracteas, aliàs maioribus, aliàs verò mi

Galileo Galilei's diagram of a telescope's optical principles, from his *Siderius Nuncius* (The Starry Messenger, 1610).

the aisle or across the aircraft. (In fact, the aircraft's engines are constantly doing work against the force of gravity, so the motion of the aircraft is not exactly uniform as in Galileo's idealized moving ship.)

Galileo deserves his immortality among physicists, as does Kepler. Yet neither scientist could offer a fundamental explanation of their most important observations. Why were the planetary orbits ellipses and not circles? And why did a projectile when fired into the air follow a parabolic trajectory, rather than, say, an elliptical one?

The answers to these questions were supplied by Newton—in his laws of motion combined with his theory of gravitation, published in his *Principia Mathematica* in 1687. This revolutionary work succeeded in unifying the motions of planetary and earthly bodies with one set of equations which could predict, given any body's mass, velocity and direction of motion, exactly how it would subsequently move under a known force. This Newtonian, mechanical view of the universe would dominate physics for the next two centuries. In the celebrated deterministic image of the astronomer and mathematician Pierre Simon de Laplace, Newton's laws would allow a supreme intelligence, if it were to be apprised of the positions and forces of all things in the universe at one instant, to predict the entire course of events from "the greatest bodies of the universe [to] those of the lightest atoms; nothing would be uncertain, and the future, like the past, would be present to its eyes."

It may seem only a short step from Galileo's discoveries to Newton's laws of motion, remarked Einstein in 1927. But, he pointed out, Galileo's mechanics were formulated to refer to a body's motion *as a whole*, while Newton's laws of motion were able to answer the question: "How does the state of motion of a mass-point change in an infinitely short time under the influence of an external force?" Newton achieved this by dissecting the trajectory of an ideal particle. In principle, by applying his laws of motion over a small difference in time he could predict the particle's position and velocity at the end of this interval; and by repeating such a calculation over and over again at successive times, he could calculate the whole trajectory. In practice, he avoided such a step-by-step calculation by inventing (in parallel with Gottfried Wilhelm Leibniz) a mathematical short cut, the differential calculus, which allowed Newton to analyse what happens to the velocity of the moving particle as the time difference becomes infinitely short: a familiar technique in mathematics. He now formulated three laws of general application to all motion whatsoever at any time in the past, present or future.

Newton's first law is simply stated. In his own words (translated from the original Latin of the *Principia Mathematica*): "Every body perseveres in its state of being at rest or of moving uniformly straight forward, except in so far as it is compelled to change its state by forces impressed." In modern language: a body continues in a state of rest or uniform motion in a straight line unless it is acted upon by external forces. This can also be called the principle of inertia, inertia being the property of matter that causes it to resist any change in its motion. Newton, like Galileo (but not Aristotle), perceived that an object that is at rest and an object that is moving uniformly can be treated by physics in the same way—an idea with profound implications and which is not at all obvious to 'everyday thinking.'

His second law was entirely his own. Again in his own words: "A change in motion is proportional to the motive force impressed and takes place along the straight line in which that force is impressed." Or in modern language: the rate of change of momentum of a moving body is proportional to and in the same direction as the force acting on it. In other words, if you push a mass twice as hard you will accelerate it (i.e. change its momentum) at twice the rate, and plainly objects try to move in the direction in which you push them, not at some angle to your push. Since the second law is one of the most famous equations in science, like Einstein's later $E=mc^2$, here it is in mathematical form:

$F=ma$

Force = mass x *acceleration.*

The proportionality factor is the mass of the body being accelerated. This equation fits with everyday experience: the heavier a bicycle is, the harder you

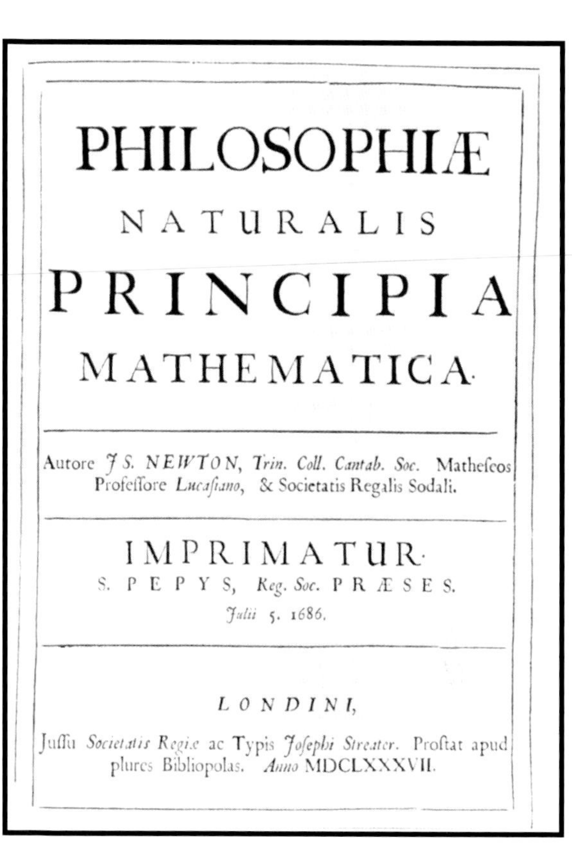

PHILOSOPHIÆ
NATURALIS
PRINCIPIA
MATHEMATICA.

Autore *J S.* NEWTON, *Trin. Coll. Cantab. Soc.* Matheſeos Profeſſore *Lucaſiano*, & Societatis Regalis Sodali.

IMPRIMATUR.
S. PEPYS, *Reg. Soc.* PRÆSES.
Julii 5. 1686.

LONDINI,
Juſſu *Societatis Regiæ* ac Typis *Joſephi Streater*. Proſtat apud plures Bibliopolas. *Anno* MDCLXXXVII.

The title page of Isaac Newton's *Principia Mathematica* (1687), published under the auspices of the Royal Society.

Nirgends in der Welt wird das Band
der Tradition und Freundschaft in
so schöner Weise gepflegt wie bei
Euch in England. So gelangt Ihr
dazu, der über-individuellen Seele
Eures Volkes eine besonders lebendige
Realität zu verleihen. Nun seid Ihr
nach Grantham gegangen um
dem grossen Genius über die trennende
Zeit hinweg die Hand zu reichen,
die Luft seiner Umgebung zu atmen,
in der er die Grundgedanken
der Mechanik, ja der physikalischen
Kausalität konzipierte. Alle,
welche ehrfürchtig über das grosse
Geheimnis des physikalischen
Geschehens nachdenken, begleiten
Euch im Geiste und teilen
das Gefühl der Bewunderung
und Liebe, das uns mit
Newton verbindet.

A. Einstein

have to push on the pedals to accelerate it to a given speed; the heavier a door is, the harder you have to push to open it; the heavier a box is, the more force you must use to lift it, so as to overcome the force due to gravity. Since the rate of acceleration of any object due to gravity is a constant at any given position on the Earth (at the surface it is equal to about 32 feet/ 9.8 metres per second squared), the second law also explains why Galileo's cannon balls fell at the same velocity from his tower, despite their different masses.

Newton's third law is somewhat counter-intuitive. It makes clear that while you sit in a chair, the chair exerts an upward force on you to balance your weight pushing down on it. And the same is true in the heavens, said Newton: while the Earth exerts a gravitational tug on the Moon, keeping it in orbit, the Moon, for its part, tugs on the Earth, creating the tides in the oceans. In the words of the *Principia*: "To any action there is always an opposite and equal reaction; in other words, the actions of two bodies upon each other are always equal and always opposite in direction."

Isaac Newton, with whom Einstein is often compared, 1690. Newton was then in his late forties.

Opposite: A draft of Einstein's letter to the Royal Society on the occasion of the bicentenary of Newton's death, 1927.

The calculation of gravity was Newton's second great contribution to mechanics. As Kepler before him had postulated ellipses for the planetary orbits, so Newton postulated the existence of an invisible force acting between masses, proportional to the sizes of the masses and inversely proportional to the square of the distance between them. This second proportionality means that if two masses are moved apart, the force of gravity between them diminishes such that when they are ten times further away the force is 100 times (ten squared) smaller than the initial attraction. In the case of the Sun, which is nearly 400 times further from the Earth than the Moon is from the Earth, the inversely proportional factor diminishing the gravitational force is about 400 squared (16,000)—but this enormous diminution is compensated by the vastly greater mass of the Sun as compared to the Moon (the Sun-to-Moon mass ratio is 30,000,000:1). And so the Earth remains in orbit around the Sun.

Newton's gravitational force acted at a distance through space faster even than the speed of light, which was first measured experimentally in 1676 at about 140,000 miles per second (quite close to its cur-

rent value). This force was totally unlike the push and pull exerted on masses in physical contact with forces, as in the projectile experiments of Galileo. Such instantaneous 'action at a distance,' with no physical explanation, naturally worried Newton, but he could see no alternative to it; and he may even have been encouraged to imagine its existence by his obsessive secret study of alchemy, which easily embraces invisible occult forces, though Newton never went so far as to cite alchemy in his published justification of his laws of motion. In the *Principia* he justifies gravity as follows: "It is enough that gravity really exists and acts according to the laws that we have set forth and is sufficient to explain all the motions of the heavenly bodies and of our sea."

Another weakness in his grand structure was that it required the existence of absolute time and space. Newton again: "Absolute, true, and mathematical time, in and of itself, and of its own nature, without reference to anything external, flows uniformly... Absolute space, of its own true nature without reference to anything external, always remains homogeneous and immovable..." In other words, a passenger on a ship moves relative to the ship, the ship moves relative to the land, the Earth moves relative to the Sun—and everything physical moves relative to a universal reference frame of space-time which is 'at rest.' But what is the nature of this universal reference frame? Newton had no real answer. "God informed Newton's creed of absolute space and absolute time," writes a Newton biographer, James Gleick. And Newton must have had some doubts about the correctness of absolute time and space, for he also noted in the *Principia*: "It may be, that there is no such thing as an equable motion, whereby time may be accurately measured. It may be that there is no body really at rest, to which the places and motions of others may be referred." To the young physics student Einstein, a similar speculation would prove highly stimulating in creating his theory of relativity.

Newton's experiments in optics are almost as important as his work on the laws of motion and gravity. But he was unable fully to understand the nature of light, preferring to regard it more as a stream of corpuscles than as a wave.

Nevertheless, looking back to his student days from his sixties, Einstein described the Newtonian basis of physics at the end of the nineteenth century as "eminently fruitful" and "regarded as final":

> *It not only gave results for the movements of the heavenly bodies, down to the most minute details, but also furnished a theory of the mechanics of discrete and continuous masses, a simple explanation of the principle of the conservation of energy and a complete and brilliant theory of heat. The explanation of the facts of electrodynamics [the physics of moving electric charges] within the Newtonian system was more forced; least convincing of all, from the very beginning, was the theory of light.*

During the nineteenth century, physicists, notably Amedeo Avogadro, James Clerk Maxwell, Ludwig Boltzmann and Josiah Willard Gibbs, had been able to apply Newton's laws of motion very successfully to gases, using the kinetic theory that gases were ensembles of constantly moving, constantly colliding atoms and molecules, even though there was as yet no incontrovertible observational evidence for the physical existence of atoms and molecules. Heating a gas was assumed to increase an atom or molecule's kinetic energy, velocity and frequency of collision with other particles, and hence it raised the pressure, temperature and rate of diffusion of the gas. The result of this model was a statistical mechanics of heat at the microscopic scale (as opposed to the scale of cannon balls or planets), known as statistical thermodynamics. This enabled physicists to calculate from first principles the observed laws of gases, such as Boyle's law connecting the pressure, volume and temperature of a gas (first formulated by Robert Boyle in Newton's day), in the same fashion that Newton's laws of motion could be used to calculate the trajectories of bodies observed by Galileo and Kepler.

The phenomena of light, and electricity and magnetism, were by contrast not so amenable to a Newtonian mechanical treatment. Newton had done brilliant work on light, which he eventually published in 1704 in his *Opticks*. His splitting and recombination of white light with prisms into the colours of the rainbow, which established that light is inherently a mixture of colours, was merely the most famous of his numerous important optical experiments. But his preference for regarding light rays as streams of particles, or 'corpuscles,' as opposed to seeing light as a wave (first proposed by Christiaan Huygens in 1678), was a serious stumbling block to Newton's comprehension of light. Reflection, refraction and diffraction could be explained by a corpuscular theory only with difficulty. For example, why, at a reflecting surface such as water, should a percentage of the light be reflected and a percentage refracted through the water? Newton postulated that some of the corpuscles were in a "fit of easy reflection" while others were in a "fit of easy transmission." This was hardly an explanation—especially as simultaneous reflection and refraction was easily explicable for waves. Though Newton knew this, and in some respects favoured a wave theory, he nevertheless put his great authority behind corpuscles, which therefore dominated the thinking of physicists long after his death in 1727. "He was justified in sticking to his corpuscular theory of light," thought Einstein, given the insecure foundation of the wave theory during Newton's age.

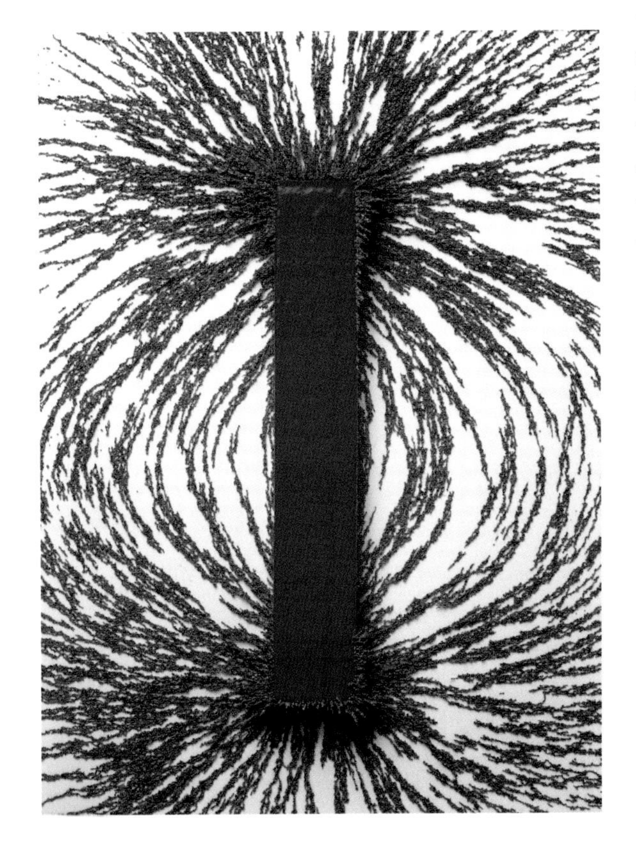

Above: Iron filings around a bar magnet demonstrate the existence of a magnetic field, a concept first proposed by Michael Faraday in the 1840s.

After 1800, however, light waves gradually took precedence over corpuscles. Thomas Young demonstrated that a light beam, when passed through two narrow double slits to create two beams, *interfered* with itself to produce a regular pattern of light and dark areas on a screen. This was an astonishing fact: light shining on light could produce more light, which was expected, but it could also produce dark, which was not at all anticipated. It had to be that the bright areas were due to the coincidence of two wave peaks and the dark ones were due to the superimposition of a wave peak and a wave trough thus cancelling the light. Further experiments on interference were done by Augustin Fresnel, who also studied examples of the polarization of light—a phenomenon possible only with waves. Fresnel concluded that light was a transverse wave which vibrated at right angles to its direction of propagation, like the ripples when a stone is dropped

James Clerk Maxwell, who was revered by Einstein, unified the phenomena of electricity and magnetism into the concept of the electromagnetic wave.

into a pond and the water itself moves *vertically* while the wave's energy spreads horizontally from the central point of impact, or somewhat like the spreading of a rumour from stationary gossip to gossip so that the rumour moves from one place to another, to use Einstein's amusing analogy. (Sound waves, by contrast, are not transverse but longitudinal waves, in which air is compressed and rarefied in the same direction as the sound propagates.) By the middle of the nineteenth century, all physicists were persuaded that light was a transverse wave.

But what was the substance that transmitted the wave energy when, for example, the Sun's light radiated through empty space to reach the Earth? This problem had not bedevilled the corpuscular theory: the corpuscles were assumed to pass through the vacuum like bullets through air. The only apparent solution was a perplexing one.

Since they believed in the mechanical view of nature, leading physicists in the second half of the nineteenth century were forced to conclude that there existed a mysterious medium, the ether, which permeated the entire universe, filling all the interstices between matter. The ether must be the transmission medium for light waves. But this entailed some contradictory properties. For various respectable physical reasons, the ether had to be "absolutely stationary, weightless, invisible, with zero viscosity, yet stronger than steel and undetectable by any instrument"—in the words of a modern theoretical physicist, Michio Kaku. Not surprisingly, during the early twentieth century, beginning with Einstein, the ether would be abandoned as an implausible concept.

Nevertheless, in the 1850s, accepting the ether, Maxwell (one of the contributors to the kinetic theory of gases) set out to establish what exactly it was that was vibrating transversely in a light wave. Maxwell's work is highly mathematical, which makes it impossible to comprehend fully without mathematics, but his most important insights and results can be summarized.

Maxwell drew on the discoveries of Faraday and Lord Kelvin about the associated phenomena of elec-

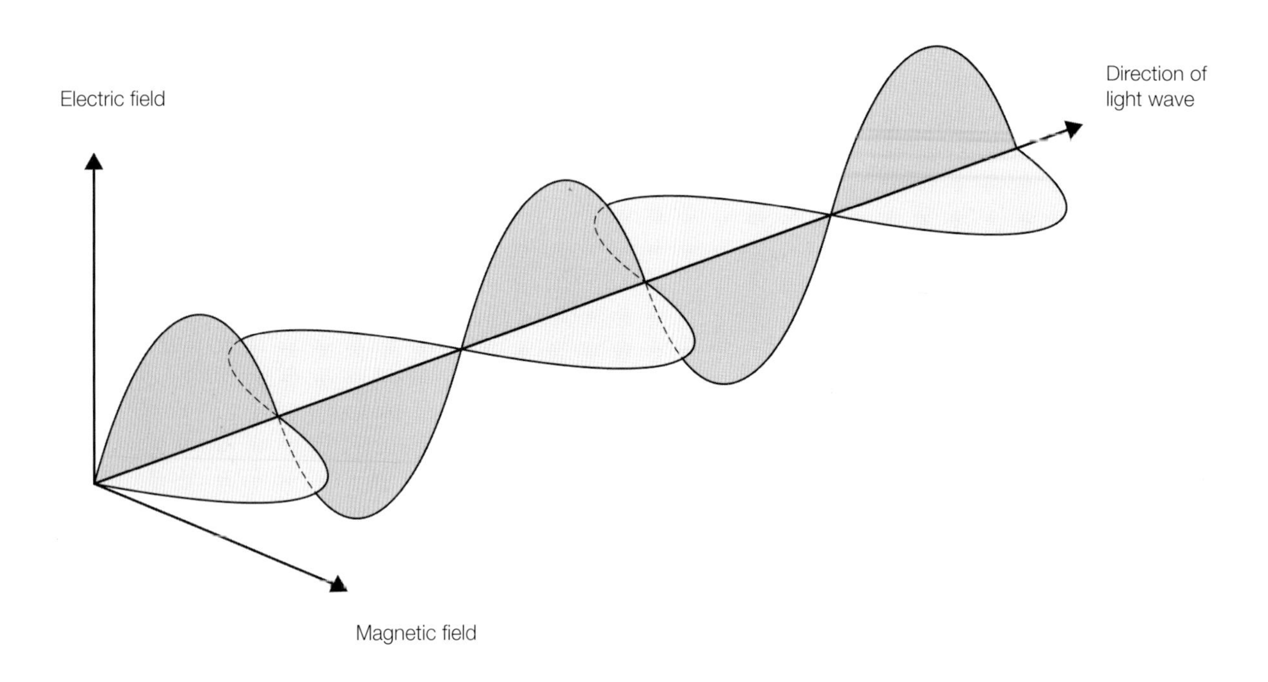

Fig. 1: A representation of an electromagnetic wave, such as light or heat. The electric and magnetic components of the wave fluctuate in planes at right angles to each other and perpendicular to the direction of propagation of the wave.

tricity and magnetism, such as the electric currents induced by moving magnets and, conversely, the magnetic fields created by electric currents, which suggested the physical reality of electric and magnetic *fields*. He derived a set of differential equations describing a new concept: an electromagnetic wave. The energy of such a wave is contained in two fields, the electric and the magnetic, which are polarized transversely at right angles to each other, while the wave itself propagates at right angles to the plane of polarization (see Fig. 1). When Maxwell calculated his wave's theoretical speed of propagation from his equations he was thrilled to discover that it coincided with the latest estimate for the speed of light from the laboratory. He thus inferred that light was probably an electromagnetic wave and published this idea in his *Treatise on Electricity and Magnetism* in 1873. A decade or so later, Heinrich Hertz began looking for Maxwell's waves. In 1888, their existence was confirmed experimentally by Hertz, who showed that radio waves, light and radiant heat were all electromagnetic waves whose behaviour was described by Maxwell's equations and that all of these waves travelled at the speed of light. Thus in electrodynamics—though not in mechanics—there was now no longer any need to take refuge in Newton's instantaneous action at a distance; the electromagnetic field transmitted electrical and magnetic forces at a finite speed, the speed of light.

On the centenary of Maxwell's birth in 1931, Einstein said:

> *Before Maxwell people conceived of physical reality...as material points, whose changes consist exclusively of motions... After Maxwell they conceived of physical reality as represented by continuous fields, not mechanically explicable... This change in the conception of reality is the most profound and fruitful one that has come to physics since Newton...*

Newton and Maxwell. They were the two most formative influences on Einstein as a fledgling physicist. Out of their ideas, and some absorbed from others, plus his own brand of 'everyday thinking,' would grow Einstein's own original contributions to physics.

Heinrich Hertz, who in 1888 provided experimental confirmation of the existence of the electromagnetic wave, as predicted by Maxwell.

Autobiographical Notes

Albert Einstein

The earliest known photograph of Einstein, aged three

Here I sit in order to write, at the age of 67, something like my own obituary. I am doing this not merely because Dr Schilpp* has persuaded me to do it; but because I do, in fact, believe that it is a good thing to show those who are striving alongside of us, how one's own striving and searching appears to one in retrospect. After some reflection, I felt how insufficient any such attempt is bound to be. For, however brief and limited one's working life may be, and however predominant may be the ways of error, the exposition of that which is worthy of communication does nonetheless not come easy—today's person of 67 is by no means the same as was the one of 50, of 30 or of 20. Every reminiscence is coloured by today's being what it is, and therefore by a deceptive point of view. This consideration could very well deter. Nevertheless much can be lifted out of one's own experience which is not open to another consciousness.

Even when I was a fairly precocious young man the nothingness of the hopes and strivings which chases most men restlessly through life came to my consciousness with considerable vitality. Moreover, I soon discovered the cruelty of that chase, which in those years was much more carefully covered up by hypocrisy and glittering words than is the case today. By the mere existence of his stomach everyone was condemned to participate in that chase. Moreover, it was possible to satisfy the stomach by such participation, but not man in so far as he is a thinking and feeling being. As the first way out there was religion, which is implanted into every child by way of the traditional education machine. Thus I came—despite the fact that I was the son of entirely irreligious (Jewish) parents—to a deep religiosity, which, however, found an abrupt ending at the age of twelve. Through the reading of popular scientific books I soon reached the conviction that much in the stories of the Bible could not be true. The consequence was a positively fanatical [orgy of] freethinking coupled with the impression that youth is intentionally being deceived by the state through lies; it was a crushing impression. Suspicion against every kind of authority grew out of this experience, a sceptical attitude towards the convictions which were alive in any specific social environment—an attitude which has never again left me, even though later on, because of a better insight into the causal connections, it lost some of its original poignancy.

It is quite clear to me that the religious paradise of youth, which was thus lost, was a first attempt to free myself from the chains of the 'merely personal,' from an existence which is dominated by wishes, hopes and primitive feelings. Out yonder there was this huge world, which exists independently of us human

*This is an extract from the opening pages of Einstein's "Autobiographical Notes," (1949) translated from the original German manuscript by Paul Arthur Schilpp.

Der Erziehungsrat

des

Kantons Aargau

urkundet hiemit:

Herr Albert Einstein von Ulm,

geboren den 14. März 1879,

besuchte die aargauische Kantonsschule & zwar die III. & IV. Klasse der Gewerbeschule.

Nach abgelegter schriftl. & mündl. Maturitätsprüfung am 18., 19. & 21. September, sowie am 30. September 1896, erhielt derselbe folgende Noten:

1. Deutsche Sprache und Litteratur	5
2. Französische " " "	3
3. Englische " " "	—
4. Italienische " " "	5
5. Geschichte	6
6. Geographie	4
7. Algebra	6
8. Geometrie (Planimetrie, Trigonometrie, Stereometrie & analytische Geometrie)	6
9. Darstellende Geometrie	6
10. Physik	6
11. Chemie	5
12. Naturgeschichte	5
* 13. Im Kunstzeichnen	4
* 14. Im technischen Zeichnen	4

* Hier gelten die Jahresleistungen

Gestützt hierauf wird demselben das Zeugnis der Reife erteilt.

Aarau den 3ten Oktober 1896.

Im Namen des Erziehungsrates,
Der Präsident:
Dr. Müri ... eli

Der Sekretar:
Frauble.

E... ung ... Noten
6, 5, ... 1.
wovon 6 die beste ... geringste ist.

Einstein's 1896 graduation certificate from the cantonal school he attended in Aarau, Switzerland. He was aged seventeen. He was awarded a "6" (the

Einstein with Paul Schilpp, the translator of his "Autobiographical Notes," from which this extract is drawn.

beings and which stands before us like a great, eternal riddle, at least partially accessible to our inspection and thinking. The contemplation of this world beckoned like a liberation, and I soon noticed that many a man whom I had learned to esteem and to admire had found inner freedom and security in devoted occupation with it. The mental grasp of this extra-personal world within the frame of the given possibilities swam as the highest aim half consciously and half unconsciously before my mind's eye. Similarly motivated men of the present and of the past, as well as the insights which they had achieved, were the friends which could not be lost. The road to this paradise was not as comfortable and alluring as the road to the religious paradise; but it has proved itself as trustworthy, and I have never regretted having chosen it.

What I have here said is true only within a certain sense, just as a drawing consisting of a few strokes can do justice to a complicated object, full of perplexing details, only in a very limited sense. If an individual enjoys well-ordered thoughts, it is quite possible that this side of his nature may grow more pronounced at the cost of other sides and thus may determine his mentality in increasing degree. In this case it is well possible that such an individual in retrospect sees a uniformly systematic development, whereas the actual experience takes place in kaleidoscopic particular situations. The manifoldness of the external situations and the narrowness of the momentary content of consciousness bring about a sort of atomizing of the life of every human being. In a man of my type the turning-point of the development lies in the fact that gradually the major interest disengages itself to a far-reaching degree from the momentary and the merely personal and turns towards the striving for a mental grasp of things. Looked at from this point of view the above schematic remarks contain as much truth as can be uttered in such brevity.

What, precisely, is 'thinking'? When, at the reception of sense-impressions, memory-pictures emerge, this is not yet 'thinking.' And when such pictures form series, each member of which calls forth another, this too is not yet 'thinking.' When, however, a certain picture turns up in many such series, then—precisely through such return—it becomes an ordering element for such series, in that it connects series which in themselves are unconnected. Such an element becomes an instrument, a concept. I think that the transition from free association or 'dreaming' to thinking is characterized by the more or less dominating role which the 'concept' plays in it. It is by no means necessary that a concept must be connected with a sensorily cognizable and reproducible sign (*word*); but when this is the case thinking becomes by means of that fact communicable.

With what right—the reader will ask—does this man operate so carelessly and primitively with ideas in such a problematic realm without making even the least effort to prove anything? My defence: all our thinking is of this nature of a free play with concepts; the justification for this play lies in the measure of survey over the experience of the senses which we are able to achieve with its aid. The concept of 'truth' cannot yet be applied to such a structure; to my thinking this concept can come in question only when a far-reaching agreement (*convention*) concerning the elements and rules of the game is already at hand.

For me it is not dubious that our thinking goes on for the most part without use of signs (words) and beyond that to a considerable degree unconsciously. For how, otherwise, should it happen that sometimes we 'wonder' quite spontaneously about some experience? This 'wondering' seems to occur when an experience comes into conflict with a world of concepts which is already sufficiently fixed in us. Whenever such a conflict is experienced hard and intensively it reacts back upon our thought world in a decisive way. The development of this thought world is in a certain sense a continuous flight from 'wonder.'

A wonder of such nature I experienced as a child of four or five years, when my father showed me a compass. That this needle behaved in such a determined

way did not at all fit into the nature of events, which could find a place in the unconscious world of concepts (effect connected with direct 'touch'). I can still remember—or at least believe I can remember—that this experience made a deep and lasting impression upon me. Something deeply hidden had to be behind things. What man sees before him from infancy causes no reaction of this kind; he is not surprised over the falling of bodies, concerning wind and rain, nor concerning the moon or about the fact that the moon does not fall down, nor concerning the differences between living and non-living matter.

At the age of twelve I experienced a second wonder of a totally different nature: in a little book dealing with Euclidian plane geometry, which came into my hands at the beginning of a school year. Here were assertions, as for example the intersection of the three altitudes of a triangle in one point, which—though by no means evident—could nevertheless be proved with such certainty that any doubt appeared to be out of the question. This lucidity and certainty made an indescribable impression upon me. That the axiom had to be accepted unproved did not disturb me. In any case it was quite sufficient for me if I could peg proofs upon propositions the validity of which did not seem to me to be dubious. For example I remember that an uncle told me the Pythagorean theorem before the holy geometry booklet had come into my hands. After much effort I succeeded in 'proving' this theorem on the basis of the similarity of triangles; in doing so it seemed to me 'evident' that the relations of the sides of the right-angled triangles would have to be completely determined by one of the acute angles. Only something which did not in similar fashion seem to be 'evident' appeared to me to be in need of any proof at all. Also, the objects with which geometry deals seemed to be of no different type than the objects of sensory perception, 'which can be seen and touched.' This primitive idea, which probably also lies at the bottom of the well-known Kantian problematic concerning the possibility of 'synthetic judgments *a priori*,' rests obviously upon the fact that the relation of geometrical concepts to objects of direct experience (rigid rod, finite interval, etc.) was unconsciously present.

If thus it appeared that it was possible to get certain knowledge of the objects of experience by means of pure thinking, this 'wonder' rested upon an error. Nevertheless, for anyone who experiences it for the first time, it is marvellous enough that man is capable at all to reach such a degree of certainty and purity in pure thinking as the Greeks showed us for the first time to be possible in geometry.

Einstein, aged five, with his sister Maja, aged three, 1884. It was around this time that Einstein's father showed him a magnetic compass, which fascinated him.

2. The Making of a Physicist

"For the detective the crime is given, the problem formulated: who killed Cock Robin? The scientist must, at least in part, commit his own crime, as well as carry out the investigation."

Einstein, *The Evolution of Physics*, 1938

Einstein, aged fourteen, in a photographic studio.

There was no hint of any intellectual distinction in Einstein's family tree. Indeed Einstein himself insisted in old age that "exploration of my ancestors...leads nowhere" in explaining his particular bent. His father Hermann was an easy-going businessman who was not very successful in electrical engineering and his paternal grandfather a merchant, while his mother Pauline, who was a fine piano player but otherwise not gifted, also came from a business family which ran a profitable grain concern and was wealthy. Though both sides of the family were Jewish, neither was orthodox. The inscriptions written in Hebrew on the nineteenth-century tombstones of the Jewish cemetery where many Einsteins lie buried in Buchau, a small township in the foothills of the Swabian Alp in southern Germany, become less frequent with time and then peter out; and the first names of the male Einsteins change from biblical ones such as Abraham into German ones like Hermann. By 1900, according to one estimate, only 15 to 20 per cent of German Jews adhered to orthodoxy. Hermann and Pauline Einstein were thoroughly assimilated and non-observant Jews ("entirely irreligious," according to their son).

Nor was there much sign of distinction in Einstein as a child. Albert was born in Ulm, the city northeast of Buchau, on 14 March 1879, the first of two children, the second being a daughter, known as Maja. He was a quiet baby, so quiet that his parents became seriously concerned and consulted a doctor about his not learning to talk. But when Maja was born in November 1881, Albert is said to have asked promptly where the wheels of his new toy were. It turned out that his ambition was to speak in complete sentences: first he would try out a sentence in his head, while moving his lips, and only then repeat it aloud. The habit lasted until his seventh year or even later. The family maidservant dubbed him "stupid."

His first school was a Catholic one in Munich, where the Einstein family had shifted from Ulm in 1880. Albert was the only Jew in a class of about 70 students. But he seems to have felt anti-Semitism among the teachers only in the religious education classes, not in the rest of the school curriculum. One day the teacher brought a long nail to the lesson and told the students that with just such nails Christ had been nailed to the Cross by the Jews. Among the students, however, anti-Semitism was commonplace, and though it was not vicious, it encouraged Albert's early sense of being an outsider, a feeling that would intensify throughout adulthood into old age.

Academically he was good yet by no means a prodigy, both at this school and at the Luitpold Gymnasium, his high school in Munich from the age of nine and a half. But Einstein showed hardly any affection for his schooling up to and including his time at the Gymnasium and in later life he excoriated the system of formal education in the Germany of his day. He disliked games and physical training, and detested anything that smacked of the military discipline typical of the Prussian ethos of northern Germany. In 1920 he even told a Berlin interviewer that the school matriculation exam should be abolished. "Let us return to Nature,

Einstein's parents Pauline and Hermann.

which upholds the principle of getting the maximum amount of effect from the minimum of effort, whereas the matriculation test does exactly the opposite."

Part of his problem lay in the heavy emphasis in the German Gymnasiums—as in British 'public' schools of the period—on the humanities; that is, on classical studies and to a lesser extent German history and literature, to the detriment of modern foreign languages. (This meant that Einstein's command of French was not fluent and that he would never be confident in speaking English, still less in writing it; nor did he learn Hebrew, to his later regret.) Science and mathematics were regarded in the Gymnasiums as the subjects with the lowest status.

But the main problem with school was probably that Albert was a confirmed autodidact. "Private study" is a term that frequently pops up in his early letters and his adult writings on education. To the average young student, such an idea is an invitation to indiscipline—maybe a chance simply to shirk—but for Einstein, studying at his own whim was his chief means of becoming educated. His sister Maja recalled that even with noisy company around him, Albert could "lie down on the sofa, pick up a pen and paper, precariously balance an inkwell on the backrest, and engross himself in a problem so much so that the background noise stimulated rather than disturbed him."

From a relatively early age he began reading mathematics and science books simply out of curiosity; at college in Zurich he ranged very widely in his reading, including the latest scientific journals, light years beyond what was prescribed by the professors; and as an adult he never read books simply because they were said to be classics, only if they appealed to him. Maybe there is a parallel here with Newton, an eclectic reader who nevertheless does not seem to have read many of the great names of his and earlier times. "Einstein was more of an artist than a scholar; in other words, he did not clutter up his mind too much with other people's ideas," in the words of the astronomer Gerald Whitrow. His long-standing friend, the physicist and Nobel laureate Max Born, recalled that:

> *Einstein expressed over and over again the thought that one should not couple the quest for knowledge with a bread-and-butter profession, but that research should be done as a private spare-time occupation. He himself wrote the first of his great treatises while earning his living as an employee of the Swiss Patent Office in Bern... What he did not consider, however, [was that] to be able to practise science as a hobby, one has to be an Einstein.*

His first scientific experience occurred when he was a child of four or five, as Einstein mentions in his "Autobiographical Notes," extracted in this book. His father Hermann showed him a magnetic compass. Observing the determined behaviour of the needle Albert was fascinated. "I can still remember," he wrote half a century later, "—or at least believe I can remember—that this experience made a deep and lasting impression upon me. Something deeply hidden had to be behind things."

Then, when he was twelve, he experienced "a second wonder of a totally different nature" while working through a small book of Euclidian plane geometry—as had Galileo at the age of 17. The "lucidity and certainty" of the geometrical proofs, based on Euclid's ten simple axioms (e.g. a circle can be constructed when its centre, and a point on it, are given), made a second deep impression, and set Einstein thinking for the rest of his life on the true relationship between mathematical forms and the same forms found in the physical world. Hence the strong appeal to him of Kepler's discovery of the planetary ellipses. The very word geometry, he noted, was from the Greek for "earth-measuring," which implied that mathematics "owes its existence to the need which was felt of learning something about the behaviour of real objects."

At the same time, he began reading popular science books brought for him by a poor medical student, Max Talmud, who would be given lunch weekly by his parents—among the few Jewish customs that the Einsteins did observe. *Kraft und Stoff* (Force and Matter) by Ludwig Büchner and the series *Naturwissenschaftliche Volksbücher* (Science for the People) by Aaron Bernstein set Einstein on course to be a scientist. They helped to prompt an intriguing thought which struck him when he was 16 and which eventually, by his own account, led to the special theory of relativity. This was

Einstein's school photograph at the Luitpold Gymnasium in Munich. He is pictured in the front row, third from the right.

the idea of chasing a beam of light and catching up with it. What would light look like if it were 'at rest' relative to an observer? Ten years later, in 1905 (as our next chapter will describe), Einstein finally accepted that such a situation was physically impossible and that Newton's concepts of absolute space and time, which apparently permitted it to occur, must be wrong.

It is ironic that a second effect of reading the science books brought by Talmud was to kill off Albert's incipient belief in orthodox religion. In the period just before, he had suddenly turned highly religious. He had stopped eating pork, started singing hymns with great fervour (and even composed a few), and begun preparing with a rabbi to become a bar mitzvah on the Sabbath following his thirteenth birthday. But the science books—although they did not attack religion as such—convinced him that much of the Bible was untrue, and induced a "suspicion against every kind of authority" which would last until his dying day.

At the Luitpold Gymnasium, things came to a head in 1894, when Albert was 15. A new class teacher informed him that "he would never get anywhere in life." When Einstein replied that surely he "had not committed any offence," he was told: "Your mere presence here undermines the class's respect for me." For the rest of his life, Einstein would be known for a mocking way with words that was sometimes biting and would always be at odds with his later gentle image. It was bound to get under the skin of authority figures—whether they were Germans, fellow Jews or, in later life, Americans. He very often mocked himself also, remarking to a friend after he became famous: "To punish me for my contempt for authority, Fate has made me an authority myself."

At home, too, all was not going well. In 1893, after a battle with larger companies, the Einstein company had failed to get a contract for lighting an important part of Munich, and the next year the decision was made to liquidate the company and set up a new one in Italy, with a new factory. Maja moved to Italy with her parents, but Albert was left alone in Munich with some distant relatives in order to take his matriculation exam. Meanwhile the beloved Einstein home was sold and quickly demolished by developers under his eyes.

The combination of disruptions at school and at home seems to have been too much for Albert, who would never refer to this unhappy period. Without consulting his parents, he got a doctor (Talmud's elder brother) to state that he was suffering from exhaustion and needed time off school, and convinced a teacher to give him a certificate of excellence in mathematics. The school authorities released him. Just after Christmas 1894, he left Munich and headed south to Milan to face his surprised parents.

Einstein did not return to the disliked Luitpold Gymnasium, and a year later rejected his German nationality too, presumably to avoid military service, becoming stateless until he was accepted as a Swiss citizen in 1901. Instead, after much private study at home in Italy in 1895 (during which he penned an immature essay on the ether concept and sent it to a maternal uncle), he sat the exam early for the Swiss Polytechnic in Zurich, probably the leading centre for the study of science in central Europe outside of Germany. He failed. However his brilliance in mathematics and physics was recognized and he was encouraged to try again the following year after further schooling. On the advice of a Polytechnic professor, he went to the Aargau cantonal school in Aarau, some 30 miles west of Zurich, which enjoyed a much less authoritarian atmosphere than the school in Munich, based as it was on the liberal ideas of the Swiss educational reformer Johann Heinrich Pestalozzi. There Albert boarded happily with the family of one of the school's teachers and started a teenage romance with a daughter of the teacher. When he took the school's final exam, which qualified him to begin study in Zurich in late 1896, he wrote a revealing essay in (execrable) French on "My plans for the future." It announced his desire to study the theoretical part of physics because of "a personal gift for abstract and mathematical thought and a lack of fantasy and practical talent," and concluded significantly: "Moreover, there is a certain independence in the profession of science that greatly appeals to me."

Switzerland became integral to Einstein's life now, during this formative intellectual period, which was also the time of his first love affair with the fellow physics student Mileva Marić who became his first wife. If there was anywhere that Einstein would have been inclined to call 'home' during his peripatetic career, it would have to be Switzerland—not his native Germany or the United States, his place of exile from Germany after 1933. In his youthful letters, his love of the Swiss mountains is transparent (though in middle age he

The Swiss Polytechnic in Zurich, where Einstein studied from 1896 to 1900.

came to prefer the infinite vistas of the ocean). It is hard to resist the feeling that the soaring, solitary splendour of the Alpine peaks must have influenced his scientific theorizing too. Later, while living in America, he wrote:

> *...creating a new theory is not like destroying an old barn and erecting a skyscraper in its place. It is rather like climbing a mountain, gaining new and wider views, discovering unexpected connections between our starting-point and its rich environment. But the point from which we started out still exists and can be seen, although it appears smaller and forms a tiny part of our broad view gained by the mastery of the obstacles on our adventurous way up.*

Thus, in the mid-1890s, Einstein would start from Newton's laws and Maxwell's equations and ascend to the heights of the field equations of general relativity 20 years later, not by overturning Newton or Maxwell but rather by subsuming them into a more comprehensive theory, somewhat as the map of a continent subsumes a map of an individual country.

Our main source for Einstein's thinking in these early years in Zurich is his correspondence with Mileva, the love letters written between 1897 and the time of their marriage in 1903, that were published only in the 1980s. The letters are peppered with references to Einstein's scientific reading on an impressive range of subjects: the electrodynamics of moving bodies, the problem of the ether and the relativity principle—which are scarcely surprising—but also molecular forces, thermo-electricity, physical chemistry and the kinetic theory of gases.

Unfortunately, there is not much scientific detail in the letters, probably because Mileva avoided science in her replies. So it is hard to penetrate the evolution of

Einstein (second from left), May 1899, with friends including Marcel Grossmann (left), the mathematician who got him a job at the Swiss Patent Office and later helped him with general relativity.

Einstein's ideas. Perhaps the closest we get comes in a letter to Mileva written in the summer of 1899 (while Albert was holidaying in a Swiss hotel with his mother and sister) which says that he is rereading Hertz on the propagation of electric force:

I'm convinced more and more that the electrodynamics of moving bodies as it is presented today doesn't correspond to reality, and that it will be possible to present it in a simpler way. The introduction of the term 'ether' into theories of electricity has led to the conception of a medium whose motion can be described, without, I believe, being able to ascribe physical meaning to it.

What *is* clear, however, is Einstein's dissatisfaction with some of the science teaching at the Swiss Polytechnic. Though he pays tribute to the professors of mathematics, such as Hermann Minkowski (who would develop special relativity mathematically after 1905), he regards the physics professors as behind the times and unable to cope with challenges to their authority. Without doubt Einstein was extremely precocious among physics students of his age in Zurich—including Mileva Marić, who was the sole woman—yet it is an astonishing fact that no course was offered at the Polytechnic on Maxwell's three-decades-old equations. Despite Hertz's experimental justification of Maxwell's equations, the electromagnetic field was still considered too recent and controversial an idea for the students.

After four years' study, most of it 'private,' Einstein graduated in the summer of 1900 with a diploma entitling him to teach mathematics in Swiss schools. His real aim was to become an assistant to a professor at the Polytechnic, write a doctoral thesis and enter the academic world as a physicist. But now his "impudence"—"my guardian angel"—of which he had made little secret, told against him.

The next two years would be very tough indeed for Albert and Mileva (who had failed to acquire a diploma). He was not offered an assistantship in Zurich, unlike some other students. Nevertheless he thought continually about physics and began to publish theoretical papers in a well-known physics journal, *Annalen der Physik*; he also wrote a thesis, which was not accepted

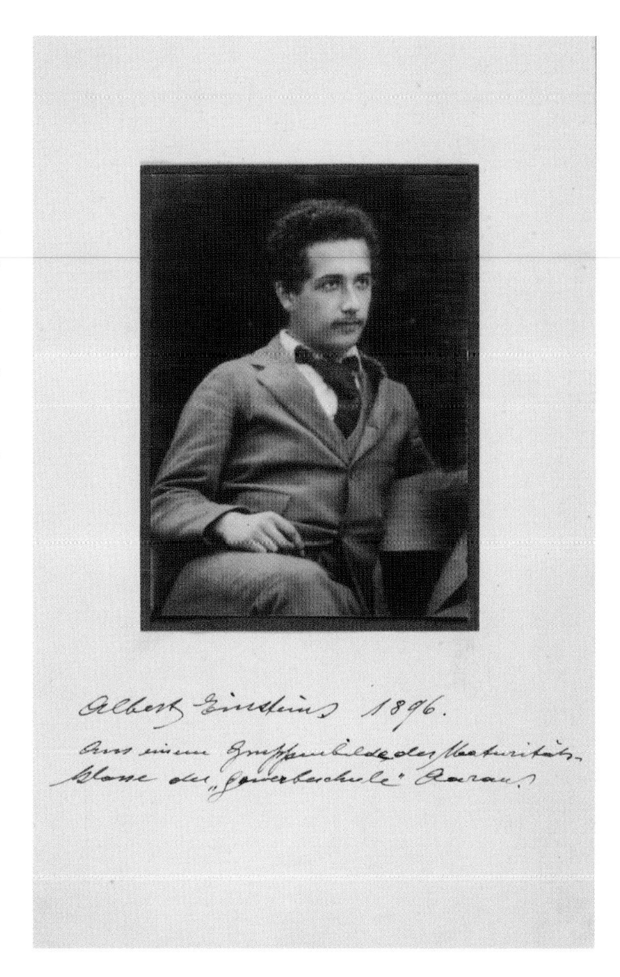

Einstein, aged 17, in a cropped version of his class photograph in Switzerland, which he has annotated: "Albert Einstein 1896. From a group picture of the graduating class of the 'vocational school' in Aarau."

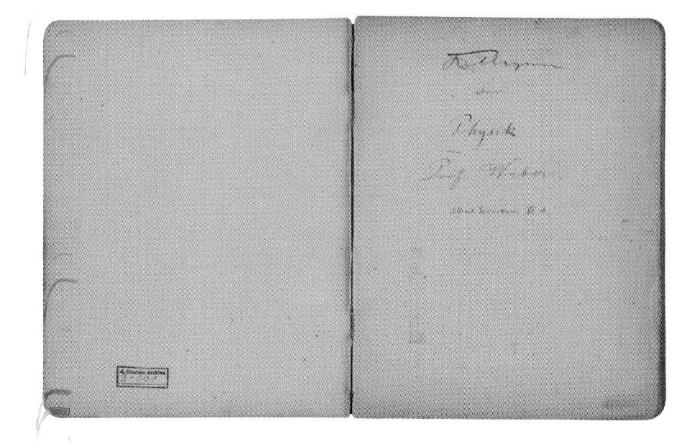

A notebook used by Einstein during Heinrich Weber's physics lectures at the Swiss Polytechnic, circa 1896–1900.

by Zurich University. But he was still an unknown and so when he wrote to notable professors offering his services he was ignored. (One of them, the chemist Wilhelm Ostwald, ironically would be the first scientist to nominate him for a Nobel prize, a mere nine years later!) "I will have soon graced all the physicists from the North Sea to the southern tip of Italy with my offer!" Albert joked wryly to Mileva in April 1901. Soon he was virtually starving, dependent on casual school teaching, and at risk of malnutrition. Then Mileva became pregnant, failed the Polytechnic exam again, and gave birth to a girl, probably named Lieserl, which had to be hushed up. (To this day, no one knows what happened to her.) Einstein's parents had always hotly opposed his proposed marriage and refused their consent; it was not given until his father lay dying in 1902, his business in bankruptcy, though his mother would never reconcile herself to the marriage. Only Einstein's unshakeable confidence in his own scientific prowess, encouraged by Mileva's single-minded devotion, could have carried him through these desperate two years.

Rescue came in the end from another fellow student at Zurich. Marcel Grossmann, who would later play a significant role in the mathematics of general relativity, secured Einstein a job at the Federal Swiss Office for Intellectual Property—the Patent Office—in Bern. Grossmann's father was a friend of the office's long-standing director, who was looking for a patent examiner with the ability to understand inventions in the burgeoning electrical industry. Einstein's knowledge of electromagnetic theory, and his considerable practical exposure to electrical devices through the family business, was deemed sufficient. On 23 June 1902, he reported for duty as a 'technical expert, third class'—the most junior post of its kind. The Swiss Patent Office would become the somewhat unlikely setting that would allow him to make his name as a physicist.

Einstein, aged 19, while at the Swiss Polytechnic in Zurich. This must have been a favourite portrait of either Einstein himself or his first wife Mileva, for it can be seen in the small picture frame on the wall behind the couple in the photograph opposite, which was taken at their home in Bern in 1904, soon after the birth of their first son Hans Albert (also pictured).

A Brief History of Relativity

Stephen Hawking

Einstein at the time he worked at the Patent Office in Bern, 1905. Opposite is a reconstruction of his Patent Office desk, 1970.

Towards the end of the nineteenth century, scientists believed they were close to a complete description of the universe. They imagined that space was filled by a continuous medium called the 'ether.' Light rays and radio signals were waves in this ether, just as sound is pressure waves in air. All that was needed for a complete theory were careful measurements of the elastic properties of the ether. In fact, anticipating such measurements, the Jefferson Lab at Harvard University was built entirely without nails so as not to interfere with delicate magnetic measurements. However, the planners forgot that the reddish-brown bricks of which the lab and most of Harvard are built contain large amounts of iron. The building is still in use today, although Harvard is still not sure how much weight a library floor without iron nails will support.

By the century's end, discrepancies in the idea of an all-pervading ether began to appear. It was expected that light would travel at a fixed speed through the ether but that if you were travelling through the ether in the same direction as the light, its speed would appear lower, and if you were travelling in the opposite direction to the light, its speed would appear higher.

Yet a series of experiments failed to support this idea. The most careful and accurate of these experiments was carried out by Albert Michelson and Edward Morley at the Case School of Applied Science in Cleveland, Ohio, in 1887. They compared the speed of light in two beams at right angles to each other. As the Earth rotates on its axis and orbits the Sun, the apparatus moves through the ether with varying speed and direction. But Michelson and Morley found no daily or yearly differences between the two beams of light. It was as if light always travelled at the same speed relative to where one was, no matter how fast and in which direction one was moving.

Based on the Michelson-Morley experiment, the Irish physicist George FitzGerald and the Dutch physicist Hendrik Lorentz suggested that bodies moving through the ether would contract and that clocks would slow down. This contraction and the slowing down of clocks would be such that people would all measure the same speed for light, no matter how they were moving with respect to the ether. (FitzGerald and Lorentz still regarded ether as a real substance.) However, in a paper written in June 1905, Einstein pointed out that if one could not detect whether or not one was moving through space, the notion of an ether was redundant. Instead, he started from the postulate that the laws of science should appear the same to all freely moving observers. In particular, they should all measure the same speed for light, no matter how fast

Albert Michelson.

Edward Morley.

they were moving. The speed of light is independent of their motion and is the same in all directions.

This required abandoning the idea that there is a universal quantity called time that all clocks would measure. Instead, everyone would have his or her own personal time. The times of two people would agree if the people were at rest with respect to each other, but not if they were moving.

This has been confirmed by a number of experiments, including one in which two accurate clocks were flown in opposite directions around the world and returned showing very slightly different times. This might suggest that if one wanted to live longer, one should keep flying to the east so that the plane's speed is added to the earth's rotation. However, the tiny fraction of a second one would gain would be more than cancelled by eating airline meals.

Einstein's postulate that the laws of nature should appear the same to all freely moving observers was the foundation of the theory of relativity, so called because it implied that only relative motion was important. Its beauty and simplicity convinced many thinkers, but there remained a lot of opposition. Einstein had overthrown two of the absolutes of nineteenth-century science: absolute rest, as represented by the ether, and absolute or universal time that all clocks would measure. Many people found this an unsettling concept. Did

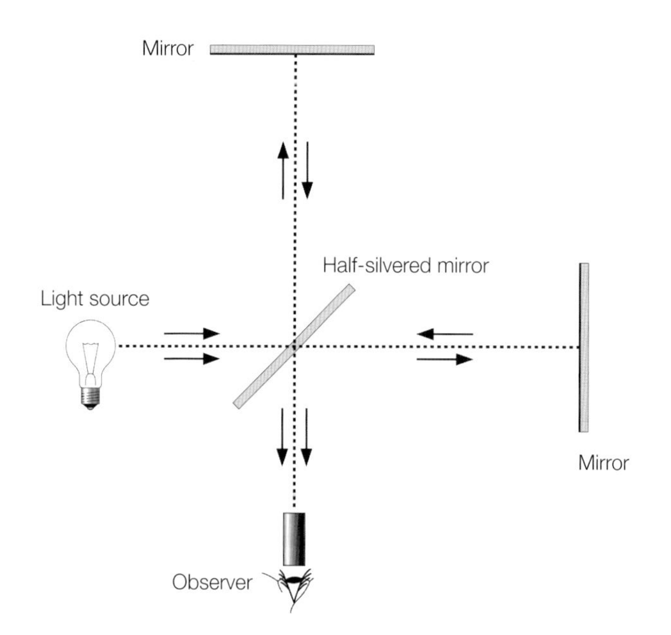

Fig 2. A simplified diagram of the Michelson-Morley experiment of 1887. Light from a source was split into two beams, at right angles to each other, by a half-silvered mirror. The two beams were then reflected by two mirrors and recombined at the half-silvered mirror into one beam, which entered the observer's telescope. If the speed of light really was affected by the supposed 'drift' of the ether through space, the two beams after reflection should have slightly different speeds and their wave crests and troughs should interfere with each other. No such interference was ever detected.

it imply, they asked, that *everything* was relative, that there were no absolute moral standards? This unease continued throughout the 1920s and 1930s. When Einstein was awarded the Nobel prize for 1921, the citation was for important but (by his standard) comparatively minor work also carried out in 1905. It made no mention of relativity, which was considered too controversial. (I still get two or three letters a week telling me Einstein was wrong.) Nevertheless, the theory of relativity is now completely accepted by the scientific community, and its predictions have been verified in countless applications.

A very important consequence of relativity is the relation between mass and energy. Einstein's postulate that the speed of light should appear the same to everyone implied that nothing could be moving faster than light. What happens is that as one uses energy to accelerate anything, whether a particle or a spaceship, its mass increases, making it harder to accelerate it further. To accelerate a particle to the speed of light would be impossible because it would take an infinite amount of energy. Mass and energy are equivalent, as is summed up in Einstein's famous equation $E = mc^2$. This is probably the only equation in physics to have recognition on the street. Among its consequences was the realization that if the nucleus of a uranium atom fissions into two nuclei with slightly less total mass, this will release a tremendous amount of energy.

In 1939, as the prospect of another world war loomed, a group of scientists who realized these implications persuaded Einstein to overcome his pacifist scruples and add his authority to a letter to President Roosevelt urging the United States to start a program of nuclear research.

This led to the Manhattan Project and ultimately to the bombs that exploded over Hiroshima and Nagasaki in 1945. Some people have blamed the atomic bomb on Einstein because he discovered the relationship between mass and energy, but that is like blaming Newton for causing aeroplanes to crash because he discovered gravity. Einstein himself took no part in the Manhattan Project and was horrified by the dropping of the bomb.

Although the theory of relativity fitted well with the laws that governed electricity and magnetism, it was not compatible with Newton's law of gravity. This law said that if one changed the distribution of matter in one region in space, the change in the gravitational field would be felt instantaneously everywhere else in the universe. Not only would this mean one could send signals faster than light (something that was forbidden by relativity); in order to know what instantaneous meant, it also required the existence of absolute or universal time, which relativity had abolished in favour of personal time.

Einstein was aware of this difficulty in 1907, while he was still at the Patent Office in Bern, but it was not until he was in Prague in 1911 that he began to think seriously about the problem. He realized that there was a close relationship between acceleration and a gravitational field. Someone inside a closed box, such as an elevator, could not tell whether the box was at rest in the Earth's gravitational field or was being accelerated by a rocket in free space. (Of course, this was before the age of *Star Trek*, and so Einstein thought of people in elevators rather than spaceships.) But one cannot accelerate or fall freely very far in an elevator before disaster strikes.

If the Earth were flat, one could equally well say that the apple fell on Newton's head because of gravity or because Newton and the surface of the Earth were accelerating upwards. This equivalence between acceleration and gravity did not seem to work for a round Earth, however—people on the opposite sides of the world would have to be accelerating in opposite directions but staying at a constant distance from each other.

But on his return to Zurich in 1912 Einstein had the brain wave of realizing that the equivalence would work if the geometry of space-time was curved and not flat, as had been assumed hitherto. His idea was that mass and energy would warp space-time in some manner yet to be determined. Objects such as apples or planets would try to move in straight lines through space-time, but their paths would appear to be bent by a gravitational field because space-time is curved.

With the help of his friend Marcel Grossmann, Einstein studied the theory of curved spaces and surfaces that had been developed earlier by Bernhard Riemann. However, Riemann thought only of space being curved. It took Einstein to realize that it is space-time which is curved. Einstein and Grossmann wrote a joint paper in 1913 in which they put forward the idea that what we think of as gravitational forces are just an expression of the fact that space-time is curved. However, because of a mistake by Einstein (who was quite human and fallible), they were not able to find the equations that related the curvature of space-time to the mass and energy in it. Einstein continued to work on the problem in Berlin, largely unaffected by the war, until he finally found the right equations in November 1915. He had discussed his ideas with the mathematician David Hilbert during a visit to the University of Göttingen in the summer of 1915, and Hilbert independently found the same equations a

Opposite: Einstein's handwritten draft of an article in *The Times*, entitled "What is the theory of relativity?", which was published on 28 November 1919. On this first page he begins by expressing his "joy and gratitude towards the astronomers and physicists of England" for their efforts in testing the implications of "a theory which was perfected and published during the war in the land of your enemies."

Was ist Relativitäts-Theorie?

Dem Ersuchen Ihres Mitarbeiters, für die „Times" etwas über „Relativität" zu schreiben, komme ich gerne nach. Denn nach dem beklagenswerten Zusammenbruch der früheren regen internationalen Beziehungen der Gelehrten ist mir dies eine willkommene Gelegenheit, meinem Gefühl der Freude und Dankbarkeit den englischen Astronomen und Physikern gegenüber auszusprechen. Es entspricht ganz den grossen und stolzen Traditionen der wissenschaftlichen Arbeit in Ihrem Lande, dass bedeutende Forscher viel Zeit und Mühe, Ihre wissenschaftlichen Institute grosse materielle Mittel aufwendeten, um eine Folgerung einer Theorie zu prüfen, die im Lande Ihrer Feinde während des Krieges vollendet und publiziert worden ist. Wenn es sich bei der Untersuchung des Einflusses des Gravitationsfeldes der Sonne auf die Lichtstrahlen auch um eine rein objektive Angelegenheit handelte, so drängt es mich doch, den englischen Fachgenossen meinen persönlichen Dank für ihr Werk zu sagen; denn ohne dasselbe hätte ich die Prüfung der wichtigsten Konsequenz meiner Theorie wohl nicht mehr erlebt. —

Man kann in der Physik Theorien verschiedener Art unterscheiden. Die meisten sind konstruktive Theorien. Diese suchen aus einem relativ few days before Einstein. Nevertheless, as Hilbert himself admitted, the credit for the new theory belonged to Einstein. It was his idea to relate gravity to the warping of space-time. It is a tribute to the civilized state of Germany at this period that such scientific discussions and exchanges could go on undisturbed even in wartime. It was a sharp contrast to the Nazi era 20 years later.

The new theory of curved space-time was called general relativity to distinguish it from the original theory without gravity, which was now known as special relativity. It was confirmed in a spectacular fashion in 1919 when a British expedition to West Africa observed a slight bending of light from a star passing near the Sun during an eclipse. Here was direct evidence that space and time are warped, and it spurred the greatest change in our perception of the universe in which we live since Euclid wrote his *Elements of Geometry* around 300 BC.

Einstein's general theory of relativity transformed space and time from a passive background in which events take place to active participants in the dynamics of the universe. This led to a great problem that remains at the forefront of physics in the twenty-first century. The universe is full of matter, and matter warps space-time in such a way that bodies fall together. Einstein found that his equations did not have a solution that described a static universe, unchanging in time. Rather than give up such an everlasting universe, which he and most other people believed in, he fudged the equations by adding a term called the cosmological constant, which warped space-time in the opposite sense, so that bodies move apart. The repulsive effect of the cosmological constant could balance the attractive effect of the matter, thus allowing a static solution for the universe. This was one of the great missed opportunities of theoretical physics. If Einstein had stuck with his original equations, he could have predicted that the universe must be either expanding or contracting. As it was, the possibility of a time-dependent universe was not taken seriously until observations in the 1920s by the 100-inch telescope on Mount Wilson.

These observations revealed that the further other galaxies are from us, the faster they are moving away. The universe is expanding, with the distance between any two galaxies steadily increasing with time. This discovery removed the need for a cosmological constant in order to have a static solution for the universe. Einstein later called the cosmological constant the greatest mistake of his life. However, it now seems that it may not have been a mistake after all: recent observations suggest that there may indeed be a small cosmological constant.

General relativity completely changed the discussion of the origin and fate of the universe. A static universe could have existed forever or could have been created in its present form at some time in the past. However, if galaxies are moving apart now, it means that they must have been closer together in the past. About fifteen billion years ago, they would all have been on top of each other and the density would have been very large. This state was called the "primeval atom" by the Catholic priest Georges Lemaître, who was the first to investigate the origin of the universe that we now call the big bang.

Einstein seems never to have taken the big bang seriously. He apparently thought that the simple model of a uniformly expanding universe would break down if one followed the motions of the galaxies back in time, and that the small sideways velocities of the galaxies might have had a previous contracting phase, with a bounce into the present expansion at a fairly moderate density. However, we now know that in order for nuclear reactions in the early universe to produce the amounts of light elements we observe around us, the density must have been ten tons per cubic inch and the temperature ten billion degrees. Further, observations of the microwave background indicate that the density was probably once a trillion trillion trillion trillion trillion trillion (1 with 72 zeros after it) tons per cubic inch. We also now know that Einstein's general theory of relativity does not allow the universe to bounce from a contracting phase to the present expansion. Roger Penrose and I were able to show that general relativity predicts that the universe began in the big bang. So Einstein's theory does imply that time has a beginning, although he was never happy with the idea.

Einstein was even more reluctant to admit that general relativity predicted that time would come to an end for massive stars when they reached the end of their life and no longer generated enough heat to balance the force of their own gravity, which was trying to make them smaller. Einstein thought that such stars would settle down to some final state, but we now know that there are no final-state configurations for stars of more than twice the mass of the Sun. Such stars will continue to shrink until they become black holes, regions of space-time that are so warped that light cannot escape from them.

Penrose and I showed that general relativity predicted that time would come to an end inside a black hole, both for the star and for any unfortunate astronaut who happened to fall into it. But both the beginning and the end of time would be places where the equations of general relativity could not be defined. Thus the theory could not predict what should emerge from the big bang. Some saw this as an indication of God's freedom to start the universe off in any way God wanted, but others (including myself) felt that the beginning of the universe should be governed by the same laws that held at other times. We have made some progress towards this goal, but we do not yet have a complete understanding of the origin of the universe.

The reason general relativity broke down at the big bang was that it was not compatible with quantum theory, the other great conceptual revolution of the early twentieth century. The first step towards quantum theory had come in 1900, when Max Planck in Berlin discovered that the radiation from a body that was glowing red hot was explainable if light could be emitted or absorbed only if it came in discrete packets, called quanta. In one of his ground-breaking papers, written in 1905, when he was at the Patent Office, Einstein showed that Planck's quantum hypothesis could explain what is called the photoelectric

effect, the way certain metals give off electrons when light falls on them. This is the basis of modern light detectors and television cameras, and it was for this work that Einstein was awarded the Nobel prize for physics.

Einstein continued to work on the quantum idea into the 1920s, but he was deeply disturbed by the work of Werner Heisenberg in Copenhagen, Paul Dirac in Cambridge, and Erwin Schrödinger in Zurich, who developed a new picture of reality called quantum mechanics. No longer did tiny particles have a definite position and speed. Instead, the more accurately one determined a particle's position, the less accurately one could determine its speed, and vice versa. Einstein was horrified by this random, unpredictable element in the basic laws and never fully accepted quantum mechanics. His feelings were expressed in his famous dictum "God does not play dice." Most other scientists, however, accepted the validity of the new quantum laws because of the explanations they gave for a whole range of previously unaccounted-for phenomena and their excellent agreement with observations. They are the basis of modern developments in chemistry, molecular biology and electronics, and the foundation for the technology that has transformed the world in the last 50 years.

In December 1932, aware that the Nazis and Hitler were about to come to power, Einstein left Germany and four months later renounced his German citizenship, spending the last 22 years of his life at the Institute for Advanced Study in Princeton, New Jersey.

In Germany, the Nazis launched a campaign against 'Jewish science' and the many German scientists who were Jews; this is part of the reason that Germany was not able to build an atomic bomb. Einstein and relativity were principal targets of this campaign. When told of the publication of a book entitled *100 Authors Against Einstein*, he replied: Why 100? If I were wrong, one would have been enough. After the Second World War, he urged the Allies to set up a world government to control the atomic bomb. In 1952, he was offered the presidency of the new state of Israel but turned it down. He once said: "Politics is for the present, while our equations are for eternity." The Einstein equations of general relativity are his best epitaph and memorial. They should last as long as the universe.

The world has changed far more in the last 100 years than in any previous century. The reason has not been new political or economic doctrines but the vast developments in technology made possible by advances in basic science. Who better symbolizes those advances than Albert Einstein?

3. The Miraculous Year, 1905

"The eternal mystery of the world is its comprehensibility… The fact that it is comprehensible is a miracle."

Einstein, "Physics and reality," 1936

The first announcement of the work that would change the course of physics occurred in late May or early June 1905, when Einstein had been working at the Swiss Patent Office in Bern for just short of three years. It came in a letter he wrote—with barely suppressed excitement—to a good friend of about the same age, Conrad Habicht, who had taken a doctorate in mathematics from Zurich University, informing him that he would soon receive a published paper in the mail, the first of four Einstein was working on. He explained to Habicht:

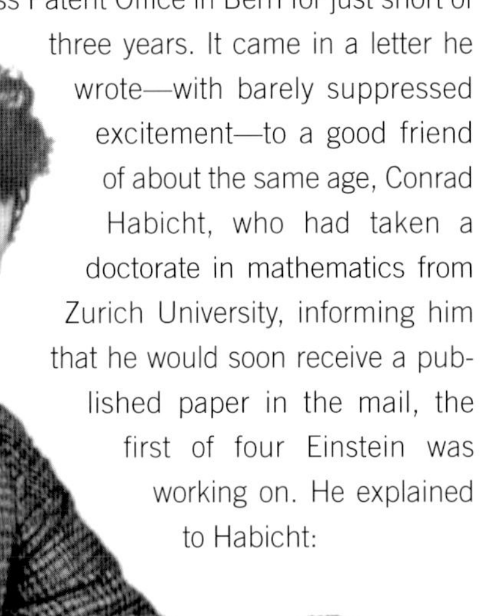

It deals with radiation and the energetic properties of light and is very revolutionary, as you will see provided you send me your paper first. The second paper is a determination of the true size of atoms by way of the diffusion and internal friction of diluted liquid solutions of neutral substances. The third proves that, on the assumption of the molecular theory of heat, particles of the order of magnitude of $^1/_{1000}$ *millimetres suspended in liquids must already perform an observable disordered movement, caused by thermal motion. Movements of small inanimate suspended bodies have in fact been observed by physiologists and called by them 'Brownian molecular movement.' The fourth paper is at the draft stage and is an electrodynamics of moving bodies, applying a modification of the theory of space and time; the purely kinematic part of this paper is certain to interest you.*

To a non-scientist, perhaps Einstein's laconic, fairly technical description does not sound all that amazing—even when one becomes aware of the fact that he would never again call any of his theories and discoveries (including general relativity) "revolutionary." But with hindsight we know that the first of these four papers published in the journal *Annalen der Physik* launched the most original scientific theory of the twentieth century, quantum theory—a theory that, unlike relativity theory, marked a distinct break with the 'classical'

Einstein at the time he worked at Patent Office in Bern, 1905.

physics of Newton and Maxwell—for which Einstein gained his Nobel prize. The second paper, which earned Einstein his doctorate from Zurich, became one of the most frequently cited papers in science, because it stimulated numerous practical applications (notably in petrochemistry). The third yielded incontrovertible proof that atoms and molecules exist and established Einstein as a founder of modern statistical thermodynamics. The fourth contained the essentials of what would soon become known as 'special' relativity, which of course led on to general relativity. All four papers were written in much less than six months in the first half of 1905—with the relativity paper, the final one, received by the journal on 30 June. And we must certainly not forget that in September 1905 Einstein added a fifth paper, a three-page coda to the relativity paper, which derived his famous equation connecting energy, mass and the speed of light, the one that would ultimately change the course of the Second World War: $E=mc^2$. All in all, 1905 was a year of miraculous scientific achievement for a virtual unknown, a patent clerk without personal contact with the leading physicists of his day, whose papers cited hardly any previous scientific authors, and who was a mere 26 years old.

No wonder, 100 years on, physicists and historians are still trying to understand how this scientific marvel transpired. There is no straightforward answer, but it is possible to hazard a few guesses.

Part of the reason for Einstein's success was surely his wide and precocious reading in science fuelled by his voracious curiosity and allied to his unusual power of concentration, all of which we have already encountered. In addition, he had an analytical ability worthy of Sherlock Holmes. A recent study by a physicist, John Rigden, remarks that Einstein "was intrigued rather than dismayed by apparent contradictions, whether they consisted of experimental results that conflicted with theoretical predictions"—as was the case with the first paper (on the quantum)—"or theories with formal inconsistencies"—as occurred with the fourth paper (on relativity).

In a related vein, two historians of science, Jürgen Renn and Robert Schulmann, point to Einstein's unwillingness to adopt received ideas simply on the authority of a scientist's high reputation—even if he

Einstein with his close friends Conrad Habicht (left) and Maurice Solovine (centre): "the Olympia Academy."

(1) k 14

Elementare Ableitung der Aequivalenz von Masse und Energie.

Die nachstehende Ableitung des Aequivalenzsatzes, die bisher nicht publiziert ist, hat zwei Vorteile. Sie bedient sich zwar des speziellen Relativitätsprinzips, setzt aber die formalen Hilfsmittel der Theorie nicht voraus, sondern bedient sich nur dreier vorbekannter Gesetzmässigkeiten:

1) des Satzes von der Erhaltung des Impulses (momentum)
2) des Ausdruckes für den Strahlungsdruck bezw. für den Impuls (momentum) eines in bestimmter Richtung sich ausbreitenden Strahlungs-Komplexes.
3) Des wohlbekannten Ausdruck für die Aberration des Lichtes (Einfluss der Bewegung der Erde auf den scheinbaren Ort der Fixsterne (Bradley).

Wir fassen nun folgendes System ins Auge. Bezüglich eines Koordinatensystems K_0 ruhend

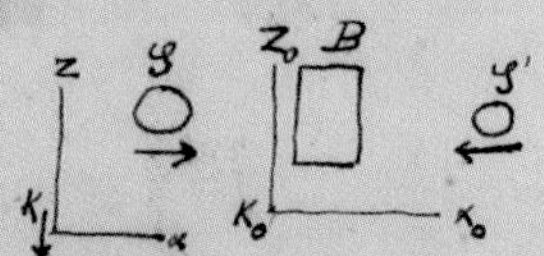

schwebe der Körper B frei im Raume. Zwei Strahlungskomplexe S, S' je von der Energie $\frac{E}{2}$ breiten sich längs der positiven bezw. negativen x_0 Axe aus und werden dann von B absorbiert. Bei der Absorption wächst die Energie von B um E. Der Körper B bleibt bei diesem Prozess aus Symmetrie-Gründen in Ruhe.

Nun betrachten wir diesen selben Prozess von einem System K aus, welches sich gegenüber K_0 mit der konstanten Geschwindigkeit v in der negativen z_0 Richtung bewegt. Inbezug auf K ist dann die Beschreibung des Vorgangs folgende

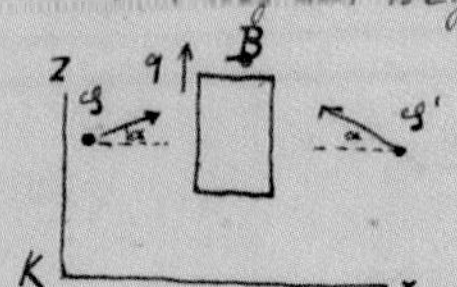

Der Körper B bewegt sich in der positiven Z-Richtung mit der Geschwindigkeit v. Die beiden Lichtkomplexe haben inbezug auf K eine Fortpflanzungsrichtung, welche einen Winkel α mit der x Axe bildet. Das Aberrationsgesetz besagt, dass in erster Näherung $\alpha = \frac{v}{c}$ ist, wobei c die Lichtgeschwindigkeit bedeutet. Aus der Betrachtung inbezug auf K_0

(2) 1.14

wissen wir, dass die Geschwindigkeit q von B durch die Absorption von S und S' keine Aenderung erfährt.

Nun wenden wir auf den Prozess inbezug auf K das Gesetz von der Erhaltung des Impulses inbezug auf die Richtung z (auf das betrachtete Gesamtsystem) an.

I. Vor der Absorption sei M die Masse von B; Mq ist dann der Ausdruck des Impulses von B (gemäss der klassischen Mechanik). Jeder der Strahlungs-Komplexe hat die Energie $\frac{E}{2}$ und deshalb gemäss einer wohlbekannten Folgerung aus Maxwells Theorie den Impuls $\frac{E}{2} \cdot \frac{1}{c}$. Dies ist streng genommen zunächst der Impuls von S inbezug auf K_0. Wenn aber q klein ist gegen c, so muss der Impuls inbezug auf K bis auf eine Grösse von der Grössenordnung ($\frac{q^2}{c^2}$ gegen 1) dieselbe sein. Von diesem Impuls fällt in die z-Richtung die Komponente $\frac{E}{2c} \sin\alpha$ oder genügend genau (bis auf Grössen höherer Ordnung) $\frac{E}{2c}\alpha$ oder $\frac{E}{2} \frac{q}{c^2}$. S und S' zusammen haben also in der z Richtung den Impuls $E\frac{q}{c^2}$. Der Gesamtimpuls des Systems vor der Absorption ist also

$$Mq + \frac{E}{c^2} q$$

II Nach der Absorption sei M' die Masse von B. Wir anticipieren hier die Möglichkeit, dass die Masse bei der Aufnahme der Energie E eine Zunahme erfahren könnte (dies ist nötig, damit das Endresultat unserer Überlegung widerspruchsfrei sei). Der Impuls des Systems nach der Absorption ist dann

$$M'q.$$

Nun setzen wir den Satz von der Erhaltung des Impulses als richtig voraus und wenden ihn inbezug auf die z Richtung an. Dies ergibt die Gleichung

$$Mq + \frac{E}{c^2} q = M'q$$

oder

$$M' - M = \frac{E}{c^2}.$$

Diese Gleichung drückt den Satz der Aequivalenz von Energie und Masse aus. Der Energie-Zuwachs E ist mit dem Massenzuwachs $\frac{E}{c^2}$ verbunden. Da die Energie ihrer üblichen Definition gemäss eine additive Konstante frei lässt, so können wir (nach passender Wahl der letzteren) statt dessen auch kürzer schreiben

$$E = Mc^2.$$

A. Einstein 1946.

The original manuscripts of Einstein's 1905 papers no longer exist, and just three documents remain which contain his famous formula, $E=mc^2$, in his own hand. This one dates from an article he wrote in 1946, for a magazine, *Science Illustrated*, under the title "$E=mc^2$: the most urgent problem of our time."

was Newton or Maxwell. For example, Einstein examined the highly influential works of Ernst Mach, a leading physicist whose *Science of Mechanics* went through 16 editions in three decades. Mach did not accept the concept of the ether or of the atom, neither of which had been physically observed, on the philosophical grounds that science should concern itself only with the summarizing of experimental results. (When Mach died in 1916, he was among the last major physicists to reject atomic theory.) Though Einstein, with his belief in the "inventions of the intellect"—such as Kepler's idea of the planetary orbits as ellipses—did not share Mach's positivist philosophy, he liked his scepticism. "[Einstein] would carefully study Mach's arguments against burdening physics with unnecessary concepts and eventually discard the ether concept, while accepting Mach's criticism of atomism as a challenge and trying to provide evidence for the existence of atoms," write Renn and Schulmann. This Einstein effectively achieved in his third paper.

Fig. 3: Brownian movement, as observed by Jean Perrin. He recorded the positions of three granules of mastic in liquid solution every 30 seconds and then connected the positions with straight lines to reveal the zigzag movement.

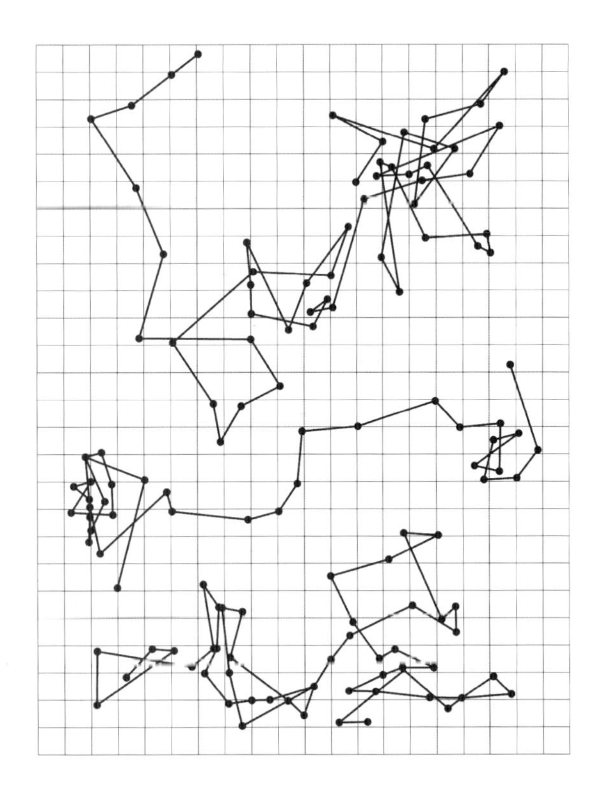

It showed how the kinetic theory of atoms and molecules could explain the long-unexplained phenomenon of Brownian movement. In 1827 the botanist Robert Brown first reported the erratic movement of very fine pollen particles suspended in water, but by Einstein's time the phenomenon had been generalized to include erratic movement wherever small inanimate particles (for instance finely ground glass) were suspended in liquids or gases, which meant that the cause had nothing to do with botany. But how could Brownian movement arise from collisions between the constantly moving atoms and molecules of the liquid or gas and stationary pollen particles? For one thing, the atoms and molecules are surely far too small to affect a comparatively huge pollen particle; the relative size of a molecule and a pollen particle is much smaller than, say, the relative size of a mosquito and an elephant. For another, the moving molecules presumably 'pin prick' a pollen particle from all directions, creating an average effect of zero. This was the view of physicists such as Boltzmann and Gibbs. Einstein, however, showed how the zigzagging of the pollen particles could arise from 'crowd behaviour' by atoms and molecules. That is, localized statistical fluctuations of large groups of atoms and molecules temporarily moving en masse first in one particular direction and then in another different direction, and at the same time pushing a pollen particle this way and that. At the end of his third paper, Einstein calculated that in water at 17 degrees Celsius particles with a diameter of a thousandth of a millimetre—that is, ten thousand times bigger than atoms—should move a mean horizontal distance of six thousandths of a millimetre in one minute. The physicist Jean Perrin soon confirmed this in the laboratory (see Fig. 3), which won him a Nobel prize in 1926. "These investigations of Einstein [did] more than any other work to convince physicists of the reality of atoms and molecules," said Max Born much later.

A third clue to Einstein's success is that he relished debate, even if his ideas got torn apart. About a year

Kramgasse 49, Bern, where the Einsteins lived from November 1903 to May 1905.

TISSOT

after his arrival in Bern in 1902, he formed a small club with Habicht and an ebullient Romanian student, Maurice Solovine, who had originally answered an advert from the impecunious Einstein offering private tuition in physics. As a joke they gave the club a high-sounding name, the Olympia Academy, with Einstein as president, and arranged to meet in the cafés of the city, at music recitals, on long walks at the weekend or in the Einsteins' small apartment. Besides reading Mach's physics and philosophy together, the "three intellectual musketeers" argued in detail about a recently published book, *Science and Hypothesis*, by the mathematician Henri Poincaré, and debated the thoughts of Hume, Spinoza and other philosophers (while also tackling some classic literature, by Sophocles, Racine and Cervantes). Sometimes Einstein would play his violin.

They also stuffed themselves with as much good food as they could afford, and generally horsed around. Once, for a special treat on Einstein's birthday, Habicht and Solovine bought some caviar. But the president got so caught up in explaining Galileo's principle of inertia that he ate it without noticing. "So that was caviar," Einstein said, when the other two finally intervened, "Well, if you offer gourmet foods to peasants like me, you know they won't appreciate it." Another time, Habicht had a tin plate engraved by a tradesman and fixed it to the Einsteins' door. It proclaimed "Albert Ritter von Steissbein, President of the Olympia Academy"—meaning roughly "Albert Knight of the Backside" or maybe something worse (since the rhyming word *Scheissbein* means 'shit-leg'!). Albert and his wife Mileva "laughed so much they thought they would die," according to Solovine. Decades later, by now laden with academic honours, Einstein remembered the Olympia Academy in a letter to Solovine as being "far less childish than those respectable ones which I later got to know." It, and discussions about science with a few other close friends in Bern, were unquestionably a key stimulus to his thinking in the period leading up to his 'miraculous year.'

The most important of all Einstein's friends at this time was probably Michele Besso (who was not an 'Olympian'), six years older than Einstein, a well-read,

Inhalt. VII

Neuntes Heft.

Ausgegeben am 22. August 1905.

Zehntes Heft.

The contents page from volume 17 of the fourth series of *Annalen der Physik*, published in Leipzig in 1905, containing Einstein's "Zur Elektrodynamik bewegter Körper" ("On the electrodynamics of moving bodies"), his special relativity paper.

quick-witted and affectionate man whose career as a mechanical engineer did not prosper because of a natural indecisiveness. Einstein had got to know him at a musical gathering in his first year as a student in Zurich and would remain in touch for six decades until Besso's death just a month before Einstein's own. In 1904, at Einstein's suggestion, Besso joined the Patent Office too, and soon the two friends were walking back and forth from the office together discussing physics. Earlier, Besso had been the one who had interested the student Einstein in Mach. Now he became the catalyst in the solving of the relativity problem.

Some time in the middle of May 1905, Einstein tells us that he went to see his friend for a chat about every aspect of relativity. After a searching discussion, Einstein returned to his apartment, and during that evening and night he saw the solution to his difficulties. The following day he went back to Besso and straightaway told him, without even saying hello: "Thank you. I've completely solved the problem. An analysis of the concept of time was my solution. Time cannot be absolutely defined, and there is an inseparable relation between time and signal velocity." His sincere gratitude can be felt in his published acknowledgement to Besso (a relative rarity with Einstein) for his "steadfastness" and for "many a valuable suggestion"—especially given the extraordinary fact that this seminal paper contains no bibliographical references at all to the works of scientists.

So how *did* Einstein come up with special relativity? Stephen Hawking, in his "Brief history of relativity" in this book, tells us that Einstein "started from the postulate that the laws of science should appear the same to all freely moving observers. In particular, they should all measure the same speed for light, no matter how fast they were moving." Let us try to unpack these tricky ideas a little.

Near the beginning of Einstein's short introduction to relativity for the general reader, he describes a

Einstein with his lifelong friend, Michele Besso (right), and Besso's wife Anna, at a reunion in Zurich, 1930.

simple but profound observation, reminiscent of Galileo's beautiful 'thought' experiment with the ship (see page 15), if more mundane. You stand at the window of a railway carriage which is travelling uniformly, in other words at constant velocity, not accelerating or decelerating—and let fall a stone on to the embankment, without throwing it. If air resistance is disregarded, you, though you are moving, see the stone descend in a straight line. But a stationary pedestrian, that is someone 'at rest,' who sees your action ("misdeed" says Einstein) from the footpath, sees the stone fall in a parabolic curve. Which of the observed paths, the straight line or the parabola, is true 'in reality,' asks Einstein? The answer is—both paths. 'Reality' here depends on which frame of reference—which system of coordinates in geometrical terms—the observer is attached to: the train's or the embankment's. One can rephrase what happens in relative terms as follows, says Einstein:

> *The stone traverses a straight line relative to a system of coordinates rigidly attached to the carriage, but relative to a system of coordinates rigidly attached to the ground (embankment) it describes a parabola. With the aid of this example it is clearly seen that there is no such thing as an independently existing trajectory (lit. 'path-curve'), but only a trajectory relative to a particular body of reference.*

Another, somewhat less familiar, situation involving relativity that bothered Einstein concerned electrodynamics. As is well known, an electric charge at rest produces no magnetic field, while a moving charge—an electric current—generates a magnetic field (circular lines of magnetic force around a current-carrying wire). Imagine a stationary electrically charged object with an observer A, also at rest relative to the object; the observer will measure no magnetic field using a compass needle. Now add an observer B moving uniformly to the east. Relative to B's reference system, the charged object (and observer A) will appear to be moving west uniformly; B, using a sensitive compass, will detect a magnetic field around the moving charged object. So, from A's point of view, there is no magnetic field around the charged object, while from B's uniformly moving point of view there *is* a magnetic field.

Anomalies of this kind intrigued Einstein. He was determined to resolve them.

It was his deeply held view that throughout the physical world the laws of mechanics, and indeed the laws of science as a whole, must be the same—'invariant' in scientific language—for all observers, whether they are 'at rest' or moving uniformly. For Einstein believed that it made no physical sense to postulate such a thing as Newton's absolute space or Maxwell's immobile ether, a universal frame of reference to which the movement of all bodies could be tacitly referred. Instead, he argued, the position in space of a body must always be specified relative to a given system of coordinates. We may choose to describe our car as moving down the motorway at a velocity of 50 miles per hour, but this figure has no absolute significance; it defines our position and speed relative only to the ground and takes no account of the Earth's rotational position and velocity around its axis or Earth's orbital position and velocity around the Sun.

But if this new postulate about the invariance of the laws of nature was actually correct, it must apply not only to moving bodies but also to electricity, magnetism and light, the electromagnetic wave of Maxwell and Hertz, which was now known from experiment to move at a constant velocity in a vacuum of about 186,000 miles per second, supposedly relative to the ether. This posed a severe problem. While Einstein was contented enough to relinquish the ether, which had never satisfied him as a concept, the constancy of the speed of light was another matter altogether.

As we know, ten years previously (maybe while preparing to take the entry exam for the Swiss Polytechnic), Einstein had reflected on what would happen if one chased light and caught up with it. He now concluded: "If I pursue a beam of light with the velocity c (velocity of light in a vacuum), I should observe such a beam of light as a spatially oscillatory electromagnetic field at rest. However, there seems to be no such thing, whether on the basis of experience or according to Maxwell's equations." To catch up with light would be as impossible as trying to see a chase scene in a movie in freeze-frame: light exists only when it moves, the chase exists only when the film's frames move through the projector. Were we to travel faster than light, Einstein imagined a situation in which we should be able to run away from a light signal and catch up with previously sent light signals. The most recently sent light signal would be detected first by our eyes, then we would see progressively older light signals. "We should catch them in a reverse order to that in which they were sent, and the train of happenings on our Earth would appear like a film shown backwards, beginning with a happy ending." The idea of catching or overtaking light was clearly absurd.

Einstein therefore formulated a radical second postulate: the speed of light is always the same in all coordinate systems, *independent* of how the emitting source or the detector moves. However fast his hypothetical vehicle might travel in chasing a beam of light, it could never catch it: relative to an observer in the vehicle the beam would always appear to travel away from him at the speed of light.

This could be true, he eventually realized, only if time, as well as space, was relative and not absolute. In order to make his first postulate about relativity compatible with his second about the speed of light, two "unjustifiable hypotheses" from classical mechanics had therefore to be abandoned. The first was that "the time-interval (time) between two events is independent of the condition of motion of the body of reference." The second was that "the space-interval (distance) between two points of a rigid body is independent of the condition of motion of the body of reference."

Thus the time of the person chasing the light wave and the time of the wave itself are not the same. Time flows for the person at a rate different from that of the wave. The faster the person goes, the slower his time flows, and therefore the less distance he covers (since distance travelled equals speed multiplied by duration of travel). As he approaches the speed of light, his watch gets slower and slower until it almost stops. In Hawking's words, relativity "required abandoning the idea that there is a universal quantity called time that all clocks would measure. Instead, everyone would have his or her own personal time." For space there is a difference, too, between the person and the light wave. The faster the person goes, the more his space contracts, and therefore the less distance he covers. As he approaches the speed of light, he shrinks to almost nothing. Depending on how close the person's speed is to the speed of light, he experiences a mixture of time slowing and space contracting, according to Einstein's equations of relativity.

These ideas seem extremely alien to us because we never travel at speeds of even a tiny fraction of the speed of light, so we never observe any 'relativistic' slowing of time or contraction of space. Our human motions seem to be governed entirely by Newton's laws (in which the speed of light is a quantity that does not even appear). Einstein himself had to struggle hard in 1905—hence his need for an intense discussion with Besso—to accept these relativistic concepts so remote from everyday experience.

With space contraction, he at least had the knowledge of a comparable earlier proposal by Lorentz and FitzGerald (mentioned by Hawking), though it had a different theoretical basis from his own and relied on the existence of the ether, a concept which Einstein had of course rejected. But the abandonment of absolute time, too, required a still greater leap of the imagination. Poincaré had questioned the concept of simultaneity in 1902 in his *Science and Hypothesis* (dissected by the Olympia Academy): "Not only do we have no direct experience of the equality of two times, but we do not even have one of the simultaneity of two events occurring in different places." Indeed Poincaré seems to have come very close to a theory of relativity just before Einstein, but apparently drew back because its implications were too disturbing to the foundations of physics. Simultaneity is a very persistent illusion for us on Earth because we so easily neglect the time of propagation of light; we think of it as 'instantaneous' relative to other familiar phenomena like sound. "We are accustomed on this account to fail to differentiate between 'simultaneously seen' and 'simultaneously happening'; and, as a result, the difference between time and local time is blurred," wrote Einstein.

Yet despite the strangeness of its predictions, relativity was built on the mechanics of Galileo and Newton modified by the electrodynamics of Maxwell, as Einstein was at pains to emphasize. Most modern physicists regard relativity theory as revolutionary, but Einstein himself did not, and reserved that adjective for his first 1905 paper on the quantum theory. Ironically, although the quantum paper was published first, his relativity paper does not refer to it; the relativity paper treats electromagnetic radiation purely as a wave and never so much as hints that it might consist of particles or quanta of energy. Presumably Einstein recognized that one big new idea per paper would be indigestible enough for most physicists. Perhaps, too, his isolating of the two ideas in two separate papers reflected his own doubts about the quantum concept. Nevertheless, with these two papers he became the first physicist to accept what is today the orthodoxy in physics: light can behave both like a wave *and* like a particle.

The mathematician Henri Poincaré, who came close to developing his own theory of relativity shortly before Einstein.

Newton, as we know, was divided about the relative merits of waves and particles, on the whole favouring the latter in his corpuscular theory of light. As for gravity, Newton had no idea at all as to how such a continuous influence might arise from discrete (in other words discontinuous) masses. In fact the debate about whether nature is fundamentally continuous or discontinuous runs through science from the atomic theory of ancient Greece right up to the present day with its opposing concepts of analogue and digital and the wave-particle 'duality' of subatomic entities like the electron. Bertrand Russell is supposed to have asked: Is the world a bucket of molasses or a pail of sand? In mathematical terms, writes a physicist (Rigden again), "Is the world to be described geometrically as endless unbroken lines, or is it to be counted with the algebra of discrete numbers? Which best describes Nature—geometry or algebra?"

Max Planck, who introduced the quantum into physics.

Quantum theory, the modern corpuscular theory, was born with the new century as a result of the work of Max Planck, though it was Einstein's paper that would endow it with its true significance. Planck considered the energy of heat that had been measured emerging from a glowing cavity, termed a 'black body' because the hole leading to the cavity behaves almost as a perfect absorber and emitter of energy with no reflecting power (like a black surface). Planck tried to devise a theory to explain how the heat energy of a black body varied over different wavelengths and at different tem-

Philipp Lenard, whose work on the generation of cathode rays may have inspired Einstein's 1905 paper on light quanta. Later, however, he became a supporter of Hitler and a persecutor of Jewish physicists, including Einstein (see Chapter 10).

peratures of the cavity. But he found that if he treated the heat as a continuous wave this wave model did not agree with experiment. Only when he assumed that the energies of the 'resonators' (atoms) in the walls of the cavity that were absorbing and emitting heat were not continuous but could take only discrete values, did theory match experiment. Instead of continuous absorption and emission of energy, energy was exchanged between heat and atoms in packets or quanta. Moreover, the size of a quantum was proportional to the frequency of the resonator, which meant that high-frequency quanta carried more energy than low-frequency quanta. As a believer in nature as a continuum, and as an innately conservative man, Planck did not feel at all comfortable with what his calculation had told him, but in 1900 he reluctantly published his theoretical explanation of black-body radiation.

Einstein was bolder than Planck. He was 20 years younger and had less stake in classical nineteenth-century physics. Probably encouraged by his disbelief in the ether (which we know had originally been introduced as a necessary medium for light waves), Einstein decided that it was not just the exchange of energy between heat/light and matter (i.e. absorption and emission) that was quantized—*light itself was quantized.* In his introduction to his first 1905 paper he stated: "According to the assumption to be contemplated here, when a light ray is spreading from a point, the energy is not distributed continuously over ever-increasing spaces, but consists of a finite number of energy quanta that are localized in points in space, move without dividing, and can be absorbed or generated as a whole." Instead of moving particles, Einstein visualized a light beam as moving packets of energy. In the 1920s, when this avant-garde concept was finally accepted by physicists, the packets were called 'photons.'

Had there been no experimental support for Einstein's "assumption" of quantized light, it would have met with an even more sceptical reaction than it in fact did. But fortunately there was some significant laboratory evidence. Though it was not detailed, Einstein audaciously interpreted the evidence with the quantum theory he had elaborated in the first part of his paper. The success of his theoretical explanation of the *photo-*

electric effect in his 1905 paper meant that his light quanta could not be totally ignored, even if they were gravely distrusted.

The photoelectric effect had been discovered by Hertz around 1888 while investigating electromagnetic waves. He noticed that in a spark gap the spark gained in brightness when illuminated by ultraviolet (high-frequency) light. With the discovery of X-rays in 1895, and of the electron in 1897, followed by the experiments of Philipp Lenard (a former assistant to Hertz), it was soon accepted that high-frequency light could knock electrons out of the surface of a metal producing photoelectrons, then called cathode rays. "I just read a wonderful paper by Lenard on the generation of cathode rays by ultra-violet light. Under the influence of this beautiful paper I am filled with such happiness and joy that I absolutely must share some of it with you," Einstein wrote to his sweetheart Mileva Marić in 1901. It may have been this paper by Lenard that started him speculating on the quantized nature of light. For Lenard's published data were in major contradiction with those expected from classical physics.

With the wave theory of light, the more intense the light, the more energy it must have and the greater the number of electrons that should be ejected from the metal. This was observed by Lenard—yet *only* above a certain frequency of light. Below this frequency threshold, no matter how intense the light, it knocked out no electrons. Moreover, above the threshold, electron emission was observed even when the light was exceedingly weak. With the quantum theory, however, Einstein realized such behaviour was to be expected. *One* quantum of light (later called a photon) would knock out *one* electron, but only if the quantum carried enough energy to extract it from the surface of the metal. Since, as Planck had shown, the size of a quantum depended on its frequency, only quanta of a sufficiently high frequency or higher would knock out electrons—hence the existence of the threshold frequency. Moreover, even a very few quanta (a very weak intensity of light) would still eject a few electrons, provided that the quanta were above the threshold in frequency.

So truly revolutionary was this discontinuous view of nature which owed almost nothing to earlier physics that light quanta took many years, much more experimentation and a lot of fresh thinking to be accepted by other physicists. Einstein was well ahead of his contemporaries in his first paper of 1905. We shall leave the quantum theory for now, and return to it in chapter five, after following the next phase in Einstein's struggle with relativity—the creation of his 'general' theory of relativity.

The spark-gap apparatus in which Heinrich Hertz discovered the photoelectric effect around 1888.

4. General Relativity

"The years of anxious searching in the dark, with their intense longing, their alternations of confidence and exhaustion and the final emergence into the light—only those who have experienced it can understand that."

Einstein, "Notes on the origin of the general theory of relativity," 1934

Unlike Einstein's quantum theory, relativity provoked a strong reaction within months of its publication in *Annalen der Physik* on 28 September 1905. German-speaking physicists were polarized by it and remarkably soon Einstein came to be held in an unusual mixture of high esteem and vigorous contempt. It was an early signal of what would happen to relativity and its creator in the world at large after the First World War—as reflected in the fact that the Swedish Academy considered the theory still too controversial even for a mention in the Nobel citation when Einstein was awarded the prize in 1922 (mainly for his first paper, the one on the quantum).

Relativity's first and strongest supporter was Max Planck, who as professor of physics at Berlin University was at the centre of German science and one of the world's leading theoretical physicists. Whatever his reservations about the quantum theory (which he had of course originated), and despite Einstein's view that relativity was a "modification" of existing work by Galileo, Newton, Maxwell and Lorentz, the cautious Planck was overwhelmed by the new theory's logic and hailed it in 1909 as unique: "In boldness it probably surpasses anything so far achieved in speculative natural science." Relativity had "brought about a revolution in our physical picture of the world, which, in extent and depth, can only be compared to that produced by the introduction of the Copernican world system." Four years later, Planck would persuade Einstein to return to his native Germany and work in Berlin.

But the very first published reaction to relativity, this time from an experimentalist, was firmly dismissive. For several years, the noted physicist Walter Kaufmann had been accelerating electrons emitted in the radioactive decay of radium (beta rays) to find out how an electron's energy increased with velocity. Today, particle accelerators are a familiar part of physics and are capable of accelerating subatomic particles to velocities close to the speed of light while measuring their increase in energy. Relativity theory predicts that at such high energies, besides increasing the velocity of an electron, the energy also goes into increasing its mass. Indeed, the mass of an electron becomes infinite at the speed of light itself. Using relativity theory the

Einstein's original manuscript in German of "The foundation of the general theory of relativity," originally published in *Annalen der Physik* in 1916. According to the Hebrew University of Jerusalem, to which he gave the 46-page manuscript on the occasion of its opening in 1925, this is probably the most valuable Einstein manuscript in existence. In this introductory paragraph, he expresses his indebtedness to the work of Minkowski and acknowledges the assistance of Grossmann.

(1)

Die Grundlage der allgemeinen Relativitätstheorie.

~~A. Prinzipielle Erwägungen zum Postulat der Relativität.~~

~~§1. Die spezielle Relativitätstheorie.~~

Die im Nachfolgenden dargelegte Theorie bildet die denkbar weitgehendste Verallgemeinerung der heute allgemein als „Relativitätstheorie" bezeichneten Theorie; die letztere nenne ich im Folgenden zur Unterscheidung von der ersteren „spezielle Relativitätstheorie" und setze sie als bekannt voraus. Die Verallgemeinerung der Relativitätstheorie wurde sehr erleichtert durch die Gestalt, welche der speziellen Relativitätstheorie durch Minkowski gegeben wurde, welcher Mathematiker zuerst die formale Gleichwertigkeit der räumlichen Koordinaten und der Zeitkoordinate klar erkannte und für den Aufbau der Theorie nutzbar machte. Die für die allgemeine Relativitätstheorie nötigen mathematischen Hilfsmittel lagen fertig bereit in dem „absoluten Differentialkalkül", welcher auf den Forschungen von Gauss, Riemann und Christoffel über nichteuklidische Mannigfaltigkeiten ruht und von Ricci und Levi-Civita in ein System gebracht und bereits auf Probleme der theoretischen Physik angewendet wurde. Ich habe im Abschnitt B der vorliegenden Abhandlung alle für uns nötigen, bei dem Physiker nicht als bekannt vorauszusetzenden mathematischen Hilfsmittel in möglichst einfacher und durchsichtiger Weise entwickelt, sodass ein Studium mathematischer Literatur für das Verständnis der vorliegenden Abhandlung nicht erforderlich ist. Endlich sei an dieser Stelle dankbar meines Freundes, des Mathematikers Grossmann gedacht, der mir durch seine Hilfe nicht nur das Studium der einschlägigen mathematischen Literatur ersparte, sondern mich auch beim Suchen nach den Feldgleichungen der Gravitation unterstützte.

A. Prinzipielle Erwägungen zum Postulat der Relativität.

§1. Bemerkungen zu der speziellen Relativitätstheorie.

Der speziellen Relativitätstheorie liegt folgendes Postulat zugrunde, welchem auch durch die Galilei-Newton'sche Mechanik Genüge geleistet wird: Wird ein Koordinatensystem K so gewählt, dass inbezug auf dasselbe die physikalischen Gesetze in ihrer einfachsten Form gelten, so gelten dieselben Gesetze auch inbezug auf jedes andere Koordinatensystem K', das relativ zu K in gleichförmiger Translationsbewegung begriffen ist. Dies Postulat nennen wir „spezielles Relativitätsprinzip". Durch das Wort „speziell" soll angedeutet werden, dass das Prinzip auf den

mass of an electron can be calculated at any given velocity, and the calculated values fit the observational data. But a century ago, such particle acceleration was in its infancy and experiments on accelerated electrons were open to different interpretations. The variation of electron mass with velocity observed by Kaufmann did not seem to agree with the predictions of special relativity. In January 1906, with deliberately dramatic emphasis, Kaufmann declared in *Annalen der Physik* that his results were "not compatible with the Lorentz-Einsteinian fundamental assumptions," and fitted better with those of two rival theories.

"I am at my wits' end," wrote Lorentz, a physicist Einstein already admired and whom he would soon come to revere. The junior patent clerk, however, though jolted, stood firm. While paying tribute to Kaufmann's carefulness, and admitting the better fit, Einstein wrote: "in my opinion, these [other] theories should be ascribed a rather small probability because their basic postulates concerning the mass of the moving electron are not made plausible by theoretical systems which encompass wider complexes of phenomena." In Einstein's view, the results of one experi-

Einstein with friends at the Swiss Polytechnic, 1913. Paul Ehrenfest, a close friend of Einstein and Hendrik Lorentz, is in the back row, fourth from right.

Opposite: Einstein in his Berlin study, 1916.

Hermann Minkowski, Einstein's mathematics professor at the Swiss Polytechnic, who went on to develop special relativity mathematically in 1908.

ment should not automatically trump a theory, if the theory can be shown to explain a range of other physical data. The best theories make many facts interlock in the structure of science.

Again and again in his subsequent career Einstein would show breathtaking confidence in the power of his theories when apparently falsified by experimental evidence—and in the vast majority of cases he would be proved right. "Subtle is the Lord, but malicious He is not," he famously remarked in 1921 on being told while touring the United States that the president of the American Physical Society, Dayton Miller, had finally proved that the elusive ether existed and modulated the speed of light—obviously a death blow to relativity, if true. Miller had refined and repeated the celebrated Michelson-Morley experiment (see pages 42–44) which in the late nineteenth century had failed to detect variation in the speed of light due to the ether. But as with Kaufmann, so with Miller, the experimenters had been deceived. Their experiments were not designed subtly enough for nature. errors were discovered in due course; and relativity was further reinforced.

While relativity certainly raised Einstein's stock in some influential circles after 1905, it took time to establish him scientifically—and indeed financially. Only after nearly four years' employment at the Swiss Patent Office was he promoted to "technical expert second class" in 1906—as a result of receiving his doctorate from Zurich (for his second, least important, paper written in 1905). Amazingly, no reference was made in the appointment process to his three path-finding papers published in the previous year. In 1907, Einstein applied to be permitted to teach at Bern University as a *Privatdozent*, submitting his published papers in place of the regulation thesis known as the *Habilitation*, and was turned down by some old-fogey professors. When, that same year, Planck's younger colleague, the brilliant Max von Laue (winner of a Nobel prize in 1914 at the age of only 35 for his work on X-ray diffraction in crystals), came to the Patent Office to make Einstein's acquaintance, Laue was so nonplussed by Einstein's youthful, modest, somewhat shabby appearance that he did not greet him and let him walk past to the waiting room. "I could not believe he could be the father of the relativity theory." On the stroll from the office to Einstein's apartment Einstein gave Laue—whom he came to admire greatly as a physicist and also as a courageous anti-Nazi—one of his favourite cheap Swiss cigars, but so horrible a smoke was it that Laue discreetly dropped it into the river!

Then in 1908 Hermann Minkowski, Einstein's former mathematics professor at Zurich, reformulated relativity mathematically and introduced the new concept of 'space-time.' At the annual meeting of German scientists in Cologne, Minkowski enthusiastically launched his idea and thereby focused physicists' attention on relativity. "The views of space and time which I wish to lay before you have sprung from the soil of experimental physics and therein lies their strength. They are radical. Henceforth space by itself, and time by itself, are doomed to fade away into mere shadows, and only a kind of union of the two will preserve an independent reality." More prosaically, events in four-dimensional space-time are analogous to points in three-dimensional space. There is an analogy too between the interval separating events in space-time and the straight-line distance between points on a flat sheet of paper. The space-time interval is absolute, in other words its value does not change with the reference frame used to compute it. In conventional space and time the stone falling from a uniformly moving train (see page 61) has two trajectories—straight down and parabolic—depending on whether it is observed from the train or from the embankment, whereas in space-time it has only *one* trajectory, which Minkowski dubbed its 'world line.' "Since the mathematicians pounced on the relativity theory I no longer understand it myself," Einstein is reported to have sighed on studying Minkowski's treatment. As a physicist he was ambivalent about mathematics, especially at this stage of his life prior to getting started on general relativity. His overall view was that "as far as the propositions of mathematics refer to reality, they are not certain; and as far as they are certain, they do not refer to reality." Even in his small book on relativity for the general reader written in 1916, he felt obliged to warn that mathematical talk of a "four-dimensional space-time continuum" had nothing at all to do with the occult or inducing "mysterious shuddering." But at the same time he did admit that without Minkowski's mathematics, the theory of general relativity might never have grown out of its nappies.

The following year, 1909, undoubtedly as a result of relativity's growing fame, Einstein's academic career took off. After seven years he left the Patent Office in Bern to become a (non-tenured) professor of theoretical physics at the university in Zurich; was the guest of honour at the next annual meeting of German scientists in Salzburg, and received his first honorary degree in Geneva at the age of just 30. In early 1911 he moved to Prague as a full professor, but stayed only 16 months before moving back to Zurich in 1912, now as full professor of theoretical physics. While based in Prague, in late 1911 he attended the first Solvay Congress in Brussels on terms of equality with the world's greatest scientists: Planck, Lorentz and Poincaré—already encountered—as well as Marie Curie, Ernest Rutherford, Walther Nernst and others. Finally, in

Einstein (second from right) at the first Solvay Congress in Brussels, 1911. It was funded by the Belgian chemist and industrialist Ernest Solvay, and organized by the chemist Walther Nernst. Participants included Marie Curie, Max Planck, Henri Poincaré and Ernest Rutherford. Einstein, then a mere 32 years old, gave the concluding address, on quantum theory.

1914, he left Switzerland and arrived in Berlin where he was elected a member of the Prussian Academy on the understanding that he could devote his entire time to research. A year and a half later, as the First World War raged, Einstein produced his more advanced theory of relativity.

At this point, after 1915, the relativity theory of 1905 begins to be known as 'special' relativity, to distinguish it from the later, more general theory. Of course the 'general' theory subsumes the 'special' theory, indeed it reduces to the special theory under conditions of uniform motion with constant velocity. In such an idealized universe, without gravity, special relativity is sufficient. But in the real physical universe, which is pervaded by gravity and accelerations due to gravity as well as various other kinds of forces, there is no such thing as absolutely uniform motion, only approximations to it, and we need the more general theory. Einstein's aim, after 1905, was to make his original relativity theory valid for *all* moving coordinate systems. Then, as he noted ironically, there would be an end to the violent disputes that had racked human thought since Copernicus, because "The two sentences, 'the Sun is at rest and the Earth moves,' or 'the Sun moves and the Earth is at rest,' would simply mean two different conventions concerning two different coordinate systems." In 1905, he had done away with Newton's concepts of absolute space and absolute time. Now, using the concept of space-time introduced by Minkowski and radically developed by Einstein with the help of his mathematician friend Marcel Grossmann, Einstein would devise a more sophisticated theory which would also do away with gravity's inexplicable instantaneous action at a distance, while at the same time retaining Newton's laws of motion and his inverse-square law of gravitational attraction as a first approximation to physical reality.

Einstein with Marie Curie, her two children and their governess, and his son Hans Albert, 1913, on an Alpine hike. Throughout the holiday, Einstein was preoccupied with his latest thoughts on gravity and relativity.

The initial inkling of how to generalize relativity struck Einstein in 1907, and it is a moment reminiscent of Newton's contemplation of the falling apple, though trickier to comprehend. "I was sitting on a chair in my patent office in Bern. Suddenly a thought struck me: if a man falls freely, he would not feel his weight." If you were to jump off a rooftop or better still a high cliff, you would not feel gravity. "I was taken aback. The simple thought experiment made a deep impression on me. It was what led me to the theory of gravity," Einstein wrote later. He called this "the happiest thought of my life."

To drive home the point, he imagined that as you fall, you let go of some rocks from your hand. What happens to them? They fall at the same rate as you, side by side. If you were to concentrate only on the rocks (admittedly difficult!) you would not be able to tell if they were falling to the ground. An observer on the ground would see you and the rocks accelerating together for a smash, but to you the rocks, relative to your reference frame, would appear to be 'at rest.'

Or imagine being inside a moving lift while standing on a weight scale. As the lift descends, the faster it accelerates, the less you will feel your weight and the lighter will be the weight reading on the scale. If the lift cable were to snap and the lift to go into free fall, your weight according to the scales would be zero. Then gravity would not exist for you in your immediate vicinity. In other words, the existence of gravity is *relative* to acceleration.

From such thinking, which became intensive only after he moved to Prague in 1911, Einstein restated a venerable idea that has become known as his 'equivalence principle'—the idea that gravity and acceleration are, in a certain sense, equivalent. It encompasses the fact, observed by Galileo, that gravity accelerates all bodies equally. In more scientific language, inertial mass (as defined by Newton's second law of motion) equals gravitational mass (as defined by gravity). Newton had simply assumed this equivalence as self-evident in formulating his gravitational equation, but Einstein felt that by understanding the physical reason for the equivalence he could gain insight into how to include gravity in relativity theory. Modern physicists have different ways to state the equivalence principle. For example, it is "the idea that the physics in an accelerated laboratory is equivalent to that in a uniform gravitational field," write Tony Hey and Patrick Walters.

For the next few years Einstein became obsessed with thoughts of accelerating closed boxes. On a Swiss Alpine hike with Marie Curie, her two daughters and their governess in the summer of 1913, Einstein toiled along crevasses and up steep rocks without seeing either, stopping periodically to discuss science. Once, Eve Curie remembered with amusement, he seized her mother's arm and burst out: "You understand, what I need to know is exactly what happens in a lift when it falls into emptiness." At a packed lecture in Vienna the following month, he entertained an audience of scientists by asking them to imagine two physicists awakening from a drugged sleep to find themselves standing in a closed box with opaque walls but with all their instruments. They would be unable to discover, he said, whether their box was at rest in the Earth's gravity or was being uniformly accelerated upwards through empty space (in which gravity is taken to be negligible) by some mysterious external force.

Max Planck giving the Planck Medal to Einstein, 1929. The first Planck Medal was given to Planck himself, and then, on the same occasion, the second medal was presented to Einstein, who declared himself unworthy of the honour.

In a similar example, Einstein imagined a small hole in the wall of a lift which is being accelerated upwards by an external force. A light ray enters the lift through the hole. The ray travels to the opposite wall of the lift. But as it does so the lift moves upwards. The ray therefore meets the opposite wall at a point a little below its point of entrance (see Fig. 4). For an observer outside the lift, there is no difficulty: the lift is accelerating upwards and so the light ray is bent downwards into a slight curve. (Had the lift been moving uniformly, the ray would have appeared to travel in a straight line.) But for an observer inside the lift who believes that the lift is at rest and that it is gravity that is acting on the lift, the curved ray poses a problem. How can a ray of light be affected by gravity? Well, said Einstein, it must be: "A beam of light carries energy and energy has mass"—as shown in his equation $E=mc^2$. "But every inertial mass is attracted by the gravitational field, as inertial and gravitational masses are equivalent. A beam of light will bend in a gravitational field exactly as a body would if thrown horizontally with a velocity equal to that of light."

The bending of light by Earth's gravity would be far too small to measure, Einstein realized. But bending might be measurable, he reasoned, when light from distant stars passed close to a massive body like the Sun. Furthermore, the equivalence principle dictated that the light emitted *from* the Sun should feel the drag of solar gravity too. Its energy must therefore fall slightly, which meant that its frequency must fall and therefore its wavelength must get longer (since light's velocity must remain constant, and the velocity of a wave equals its frequency multiplied by its wavelength). So light from atoms in the surface of the Sun, as compared with light emitted by the same atoms in interstellar space, should be shifted towards the red, longer-wavelength, end of the visible spectrum when observed on Earth. The bending of light by the Sun and the *gravitational red shift* were therefore possible tests of relativity.

Gravity Probe B, launched by NASA on 20 April 2004, is intended to test for the gravitational distortions of space-time predicted by Einstein's general theory of relativity. The mission uses four ultra-precise gyroscopes, which orbit the Earth in a unique satellite. (See page 77.)

But in order to introduce gravity into relativity, a major problem confronted Einstein in trying to apply the equivalence principle to the flat space-time visualized by Minkowski. The problem can be perceived, at least dimly, from a paradox about the simple merry-go-round which bothered Einstein. When a merry-go-round is at rest, its circumference is equal to π times its diameter. But when it spins, its circumference travels faster than its interior. According to relativity, the circumference should therefore shrink more than the interior (since space contraction increases with velocity), which must distort the shape of the merry-go-round and make the circumference less than π times the diameter. The result is that the surface is no longer flat; space is curved. Euclid's geometry, based on flat surfaces and straight light rays, no longer applies. Einstein is said to have had a nice analogy for this curvature, which he gave to his young son who once asked his father why he was so famous: "When a blind beetle crawls over the surface of a curved branch, it doesn't notice that the track it has covered is indeed curved. I was lucky enough to notice what the beetle didn't notice."

In the mid-nineteenth century, the mathematician Bernhard Riemann had invented a geometry of curved space in which, said Einstein, "space was deprived of its rigidity, and the possibility of its partaking in physical events was recognized." Now Einstein—initially with the help of the mathematician Grossmann but after he moved to Berlin in 1914 almost entirely alone—used Riemann's geometry to create a new geometry of curved space-time, as described by Stephen Hawking in this book. "His idea was that mass and energy would warp space-time in some manner yet to be determined." Gravity would no longer be an interaction of bodies through a law of forces; it would be a field effect that emerged from the way in which mass curved space. When a marble is propelled across a flat, smooth trampoline on which sits a large and heavy ball, the marble follows a curved path around the depression caused by the ball (see Fig. 5, page 77). In the Newtonian view a gravitational force emanates from the ball and somehow compels the marble to move in a curve. But according to general relativity it is the curvature of space—or rather space-time—that is responsible; there is no mysterious force. Matter tells space how to curve; space tells matter how to move—this is a grossly simplified summary of Einstein's general theory of relativity.

The light 'corpuscles' in a light ray from a distant star grazing the Sun on its way to our eyes could be

Fig. 4: Einstein's 'thought' experiment on the bending of a light ray in a lift which is accelerating upwards.

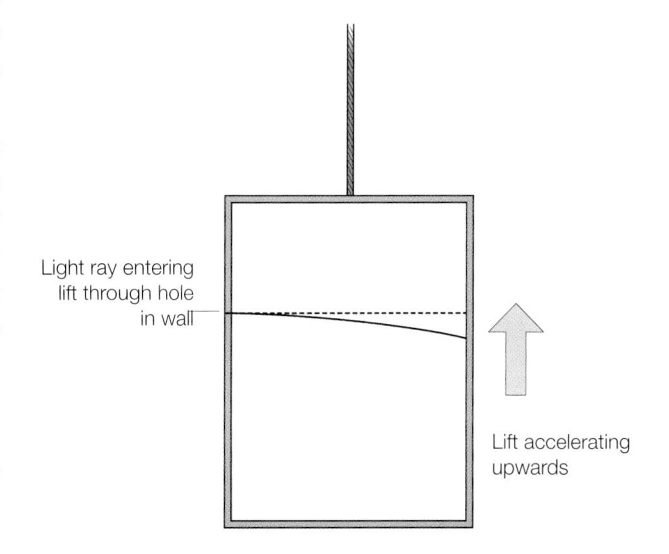

One of the coated gyroscope rotors and its matching housings, used in Gravity Probe B.

interpreted like marbles moving past a ball. In 1911, before he had mastered curved space-time, Einstein had calculated the expected bending of starlight on the basis of Newton's law of gravitation. In 1915, however, having completed general relativity, he recalculated the deflection as *twice* the size of his 1911 calculation. If the magnitude of the actual deflection were to be measured, it would test which gravitational theory was correct: Newton's or Einstein's. "The examination of the correctness or otherwise of this deduction is a problem of the greatest importance, the early solution of which is to be expected of astronomers," wrote Einstein in his *Relativity*, published in 1916. Three years later, on 29 May 1919, a solar eclipse visible in West Africa and Brazil allowed the deflection to be measured by a team led by the astronomer Arthur Eddington. Although the results were somewhat unclear, due to the difficulty of measuring the tiny angle of deflection precisely enough and to bad weather obscuring the eclipse, Eddington nevertheless declared that Einstein's theoretical prediction had been vindicated by nature. General relativity received a tremendous boost, and Einstein's world fame was assured.

The news reached Einstein in Berlin in a telegram from his friend Lorentz. He was certainly pleased (rejoicing in a postcard to his mother) but not by any means ecstatically 'over the moon.' He showed the telegram to a doctoral student who was with him and told her: "I knew all the time that the theory was correct." But supposing the result had been equivocal or contradicted his theory, the student asked? "In that case I'd have felt sorry for God, because the theory is correct." Almost 30 years later, when Planck died, Einstein told a friend, after warmly praising Planck, "but, you know, he didn't really understand physics. During the eclipse of 1919, Planck stayed up all night to see if it would confirm the bending of light by the gravitational field of the Sun. If he had really understood the way the general theory of relativity explains the equivalence of inertial and gravitational mass, he would have gone to bed the way I did." As for Eddington, the leader

A telescope and other optical equipment used to observe the 1919 solar eclipse from Sobral in northern Brazil on 29 May.

of the eclipse astronomers, he was so utterly convinced of the truth of Einstein's theory that "had he been left to himself, he would not have planned the expeditions"! (as Eddington confessed later to the young astrophysicist Subrahmanyan Chandrasekhar).

Full confirmation of general relativity took a lot longer though. The predicted gravitational red shift, for example, was not experimentally settled until the early 1960s; other tests, still more refined, continue today. For example Gravity Probe B, launched into space in 2004, carries high-precision gyroscopes to look for the tiny gravitational distortions in the fabric of space caused by the Earth that are predicted by Einstein's theory. Only with the boom in astronomy and cosmology of the last three or four decades has general relativity come to occupy a more central stage in physics, as we shall see in Chapter 7. For several decades after 1919, the subject stagnated—and was frankly distrusted by many physicists, especially in the United States where 'practical' scientists regarded relativity as a kind of German metaphysics—though Einstein himself continued to make some theoretical contributions to it. In the 1920s and 1930s the focus of attention in physics was not on relativity but on quantum mechanics and the implications of the light quanta postulated in Einstein's "revolutionary" paper of 1905. So it is back to quantum theory that we shall now go.

Einstein in Leiden with Hendrik Lorentz (middle) and Arthur Eddington (right), early 1920s. Eddington led a team of British astronomers to observe the eclipse of 1919 and from the beginning was the leading scientific champion of general relativity.

Fig. 5: A marble, projected past a heavy ball on a trampoline, simulates the distortion of space-time. This distortion explains the bending of starlight by the Sun, which is observable during a solar eclipse.

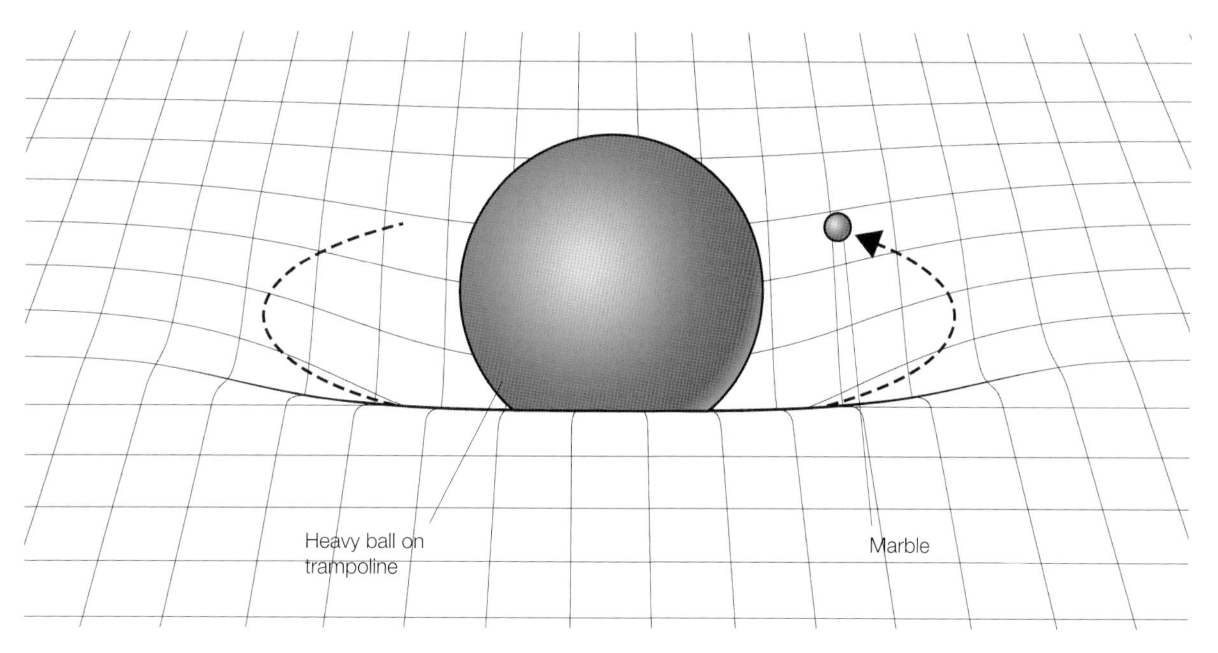

Varying c: Vodka Without Alcohol?

João Magueijo

A century after Einstein's theory of relativity was invented, almost everyone knows it enshrines the principle that the speed of light never changes. Physicists are obsessed with the idea, which takes many different forms. The speed of light is the same whether you move the light source, such as a torch, or move your eyes with respect to the torch. It is the same whether you go near the Sun (or a black hole) or instead remain in more clement environments like the Earth, well away from strong gravity. It is the same in all epochs of the universe, even at 'interesting' times, such as during the birth of the cosmos near the moment of the big bang. The speed of light is also the same throughout the rainbow, for each of its colours, that is to say for all of light's energies. It is even the same for stuff that is not exactly 'light' but which resembles it enough to share light's speed of propagation—gravity, for example. Gravity, too, supposedly moves at the speed of light.

Etcetera, etcetera, etcetera...

Such stubborn constancy has made c the reliable and certain pillar of modern physics, a safe haven where the physicist may attempt to define rigidity in a world rife with variation. Unsurprisingly, then, the idea of a varying speed of light (VSL) might sound like a perfect anathema for physicists, their equivalent of alcohol-free vodka. So you may be taken aback to learn that the first VSL theory was proposed in 1911. By Einstein himself.

Strangely, the man who created what many now regard as a dogma, was not at all dogmatic about it. His particular argument is long forgotten and does not really matter any more (suffice to say, it led him to the general theory of relativity in 1915). What counts is his attitude: as long as we have a good reason, we should be ready to give up any scientific principle, no matter how sacred.

In the twenty-first century we may be forced to adopt this broad-minded attitude again. A number of fundamental problems in physics are begging for the constancy rule to be broken. Many physicists feel that it is too early to abandon ship, and they—like the proponents of the ether concept a hundred years ago—keep up a fight that may become a contradiction in terms. Some others, including myself, are ready to embrace a varying speed of light as part of progress.

Einstein as professor of theoretical physics in Prague, 1912, a year after he first proposed that the speed of light might vary.

The VSL story so far is in two parts. The first is about the search for a theory called quantum gravity, the second involves cosmology, the study of the structure of the universe.

J. F. Langhans

Attempts to subject gravity to the rules of the quantum have failed abysmally. The problem is invariably that we want to make space and time discrete—something that does not seem to be consistent with the continuum in the theory of relativity.

Quantum theory replaces the continuum with granules, for instance light rays become shoals of quantum particles called photons. Gravity being a theory of space and time, quantum gravity ought therefore to introduce potholes into the fabric of space and substitute a necklace made up of special beads—'atoms of time,' so to speak—for the smooth flow of time. But in relativity theory, the price we pay for the constancy of the speed of light is that space contracts and time slows down, depending on the observer. If the theory of relativity is correct, and so also is the quantum theory, then the potholes and beads would have to vary from observer to observer, and no one would agree on the fabric of space-time. Such variation is acceptable if space-time is a continuum, but it is not acceptable if space is discrete. For one thing observers would disagree over the point at which discrete space 'coarse-grains'—transforms—into a continuum. Overall, a never-ending parade of contradictions results. This is the main reason why we have not succeeded in quantizing gravity.

A simple solution would be to make the speed of light colour-dependent. For light with wavelengths similar in size to the 'atoms' of space and frequencies related to the 'atoms' of time, it could be that the speed of light becomes infinite. An infinite speed would bring rigidity back into space and time (that is why the Newtonian concept of absolute space and time postulates an infinite speed of light). The quanta of space-time would become absolute, and at least some of the paradoxes of quantum gravity would be resolved.

In cosmology, the story is somewhat different. Here the issue is that the speed of light, besides enjoying a special status in physics, acts as a universal speed limit. While sitting in a traffic jam you might think such a constraint is immaterial. However on a cosmic scale it leads to a plethora of troubles.

Space travel is perhaps the best-known disaster area. Consider cosmic distances and compute the travel times if the speed of light is the universal speed limit. Compare them with the human lifetime. And despair. We humans are stuck in our little corner of the universe.

This well-publicized annoyance has its counterpart in cosmology. The constancy of the speed of light forbids the cosmologist from explaining the most obvious features of the universe: why is it so flat and so homogeneous, and where do its galaxies come from? Just after the big bang the universe had existed for a very short time, and matter could not have travelled very far, not even at the speed of light. According to the theory of relativity, the universe should therefore have broken up into a myriad of tiny disconnected horizons. Hence the cosmologist can never hope to explain the large-scale features of the universe with a constant speed of light. The contradictions can only be resolved if the very early universe was pervaded by contact, if the horizons became infinite—which requires that the speed limit of light must increase near the moment of cosmic creation. Yet again we are tempted by a VSL theory.

So we have two good reasons for committing a heresy. But is it really so heretical? The fact is, the constancy of the speed of light touches so many aspects of physics that to date only some of these aspects have been well tested by experiment. The rest of the argument against a VSL theory is just intellectual inertia.

It would be foolish to let go of the constancy of c lightly. It has served us well in the 100 years since Einstein's relativity paper of 1905. But it would be even more foolish to treat it as a dogma. Nothing in science should have that status. Everything should be challenged, even if only out of pure despair.

Einstein visits the Einstein Tower in Potsdam, near Berlin. It was opened in December 1924 as a solar observatory and was intended to test some of Einstein's theoretical predictions, including the gravitational red shift. The modernist design was by Erich Mendelsohn and was much admired; but Einstein did not appreciate it.

5. Arguing About Quantum Theory

"Quantum mechanics is certainly imposing. But an inner voice tells me that it is not yet the real thing. The theory says a lot, but does not bring us any closer to the secrets of the 'old one.' I, at any rate, am convinced that *He* is not playing at dice." Einstein, letter to Max Born, 1926

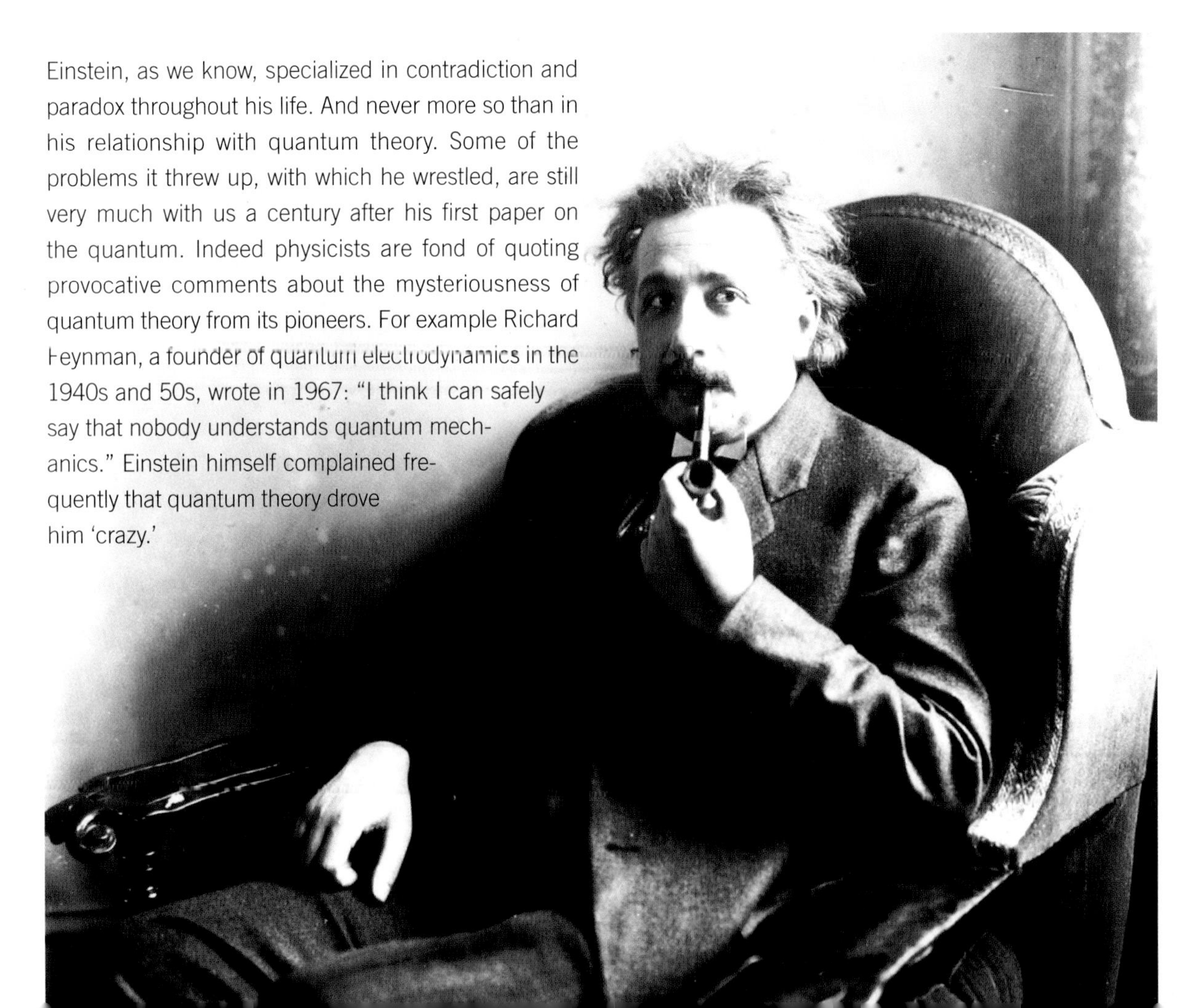

Einstein, as we know, specialized in contradiction and paradox throughout his life. And never more so than in his relationship with quantum theory. Some of the problems it threw up, with which he wrestled, are still very much with us a century after his first paper on the quantum. Indeed physicists are fond of quoting provocative comments about the mysteriousness of quantum theory from its pioneers. For example Richard Feynman, a founder of quantum electrodynamics in the 1940s and 50s, wrote in 1967: "I think I can safely say that nobody understands quantum mechanics." Einstein himself complained frequently that quantum theory drove him 'crazy.'

Rotogravure Picture Section

The New York

FIVE NOBEL PRIZE WINNERS AT A DINNER IN BERLIN: PROFESSOR ALBERT EINSTEIN,
With (at His Right) Professor Walter Nernst and (at His Left) Professor Max Planck, Dr. Robert A. Millikan and Professor Max von Laue, the Host at the Party Given for Dr. Millikan.
(Times Wide World Photos, Berlin Bureau.)

There are two main phases in his role in the quantum story. The first phase ran from 1905, when Einstein published his "revolutionary" paper assuming the existence of light quanta (photons), to the mid-1920s, when his assumption was unequivocally validated by the experiments of Arthur Compton showing that X-rays were scattered by the free electrons in metal foil according to quantum rules. During this phase, now known as the period of the 'old' quantum theory, Einstein stood almost alone in his belief that light itself was quantized, while making major contributions to physics by employing the hypothesis. (One of them would form the theoretical basis of the laser invented after his death.) This is not to say that quantum concepts were not seriously discussed then; they certainly were, as in Planck's theory of the absorption and emission of black-body radiation and Niels Bohr's solar-system model of the atom. But Einstein's light quanta were considered too radical for polite scientific society. In 1922, they were evaded in Einstein's Nobel citation, which referred instead to his "services to theoretical physics and especially for his discovery of the law of the photoelectric effect," and the same happened the following year when the prize was given to Robert Millikan partly for his experimental confirmation of the same law. During this first phase "it was the *law* that was accepted, *not* the photon," the physicist Andrew Whitaker stressed recently. "Einstein's initial conception of the photon was no less than an act of genius, and his perseverance with it, despite the negative response and his own misgivings over the relation between the wave and particle concepts, showed great determination and courage."

Einstein with Pieter Zeeman (left) in Zeeman's laboratory in Amsterdam, 1920. Zeeman's work on atomic spectra at the turn of the century was an important step towards understanding the structure of the atom. To the right of Einstein is his friend Paul Ehrenfest.

Then in 1925–26, the beginning of the second phase, along came quantum mechanics, originated by Werner Heisenberg and Erwin Schrödinger, together with Niels Bohr, Max Born, Louis de Broglie, Paul Dirac and others—but not Einstein. Once again he stood almost alone. In this phase of the quantum revolution, from 1926 until his death in 1955, Einstein was profoundly sceptical about his colleagues' new interpretation of physical reality in terms of probability, indeterminacy and uncertainty—which did not prevent it from quickly becoming the orthodoxy it remains today, not only in physics and chemistry but throughout science. In Einstein's view, however successful quantum theory might be in describing natural phenomena, it remained incomplete, an insufficiently fundamental explanation of the universe, like gravitational theory before the invention of general relativity. As Einstein wrote of Newton in 1933, "the tremendous practical success of his doctrines may well have prevented him and the physicists of the eighteenth and nineteenth centuries from recognizing the fictitious character of the foundation of the system." He believed that the same would eventually prove true of the quantum theory.

Three separate comments from three pivotal figures in twentieth-century science, each of them close to Einstein, give a fair picture of the deep resistance to his 1905 theory of quantized light in the days of the old quantum theory.

In 1910, the physical chemist Walther Nernst (soon to be a colleague of Einstein in Berlin) called Einstein's quantum hypothesis "probably the strangest thing ever thought up. If correct, it opens entirely new roads for so-called ether physics and for all molecular theories. If false, it will remain 'a beautiful memory' for all times." To Nernst, the hypothesis apparently looked alluring but illicit. The following year, he convened the first Solvay Congress in Brussels in order to discuss the implications of the quantum hypothesis, and invited Einstein to give the concluding address.

Less sympathetic was Planck, who was of course a theoretical physicist like Einstein. He had greeted relativity as the work of a new Copernicus, but he was embarrassed by the intellectual offspring of his own quantum theory of black-body radiation. Even when fulsomely recommending Einstein for membership of the Prussian Academy in 1913, Planck felt obliged to add a gentle apology for his distinguished protégé's subversive notion: "That sometimes, as for instance in his hypothesis of light quanta, he may have gone overboard in his speculations should not be held against him too much, for without occasional venture or risk no genuine innovation can be accomplished even in the most exact sciences."

Least sympathetic of all was the experimental physicist Millikan, the American who had first measured the charge on the electron in his classic oil-drop experiment. (He was later a leading light of the California Institute of Technology, to which he invited Einstein as an honoured guest in the 1930s.) In 1916, Millikan bluntly called Einstein's quantum theory "wholly untenable," a "bold, not to say reckless, hypothesis," in two published scientific papers. What is particularly pointed about this criticism is that Millikan had just spent ten years in his laboratory testing the predictions of Einstein's 1905 equation for the photoelectric effect and reluctantly confirmed its striking accuracy. Even so, Millikan refused to accept Einstein's theoretical explanation of his own experimental results, because to have allowed the existence of light quanta would have appeared to contradict absolutely the ruling wave concept of light.

The earliest of Einstein's applications of the quantum hypothesis (after the photoelectric effect), which he published in 1907, was not directly to do with light quanta. It concerned the solid state. In the 1820s Pierre Dulong and Alexis Petit had experimented with heating various metallic elements such as copper, nickel and gold, and had discovered an interesting and useful rule, the so-called Dulong and Petit law. It states that the amount of energy required to increase the temperature of one kilogram of a substance by one degree—known as its specific heat capacity—is inversely proportional to its atomic weight (more accurately, its relative atomic mass). In simple terms this means that more heat is required to raise the temperature by the same degree of one kilogram of carbon (e.g. coal)—atomic weight 12—than of one kilogram of gold—atomic weight 79. The reason is that there are more atoms of carbon in the lump of coal than there are atoms of gold in the equivalent lump of gold, and each of these atoms needs to absorb heat in order to raise the overall temperature. The greater the atomic weight of an element, the smaller its specific heat capacity; multiplied together, the atomic weight and the specific heat equalled a constant value across a number of different substances, as measured by Dulong and Petit. Their law was thus good evidence for the atomic structure of matter and also suggested the surprising fact that the atoms of a range of different elements had exactly the same capacity for heat regardless of their atomic weight.

But as Einstein had been aware since his student days, Dulong and Petit's law worked well only at high temperatures and only with certain elements, not with others. At low temperatures the specific heat capacity fell and the law was not obeyed; and for diamond (carbon), boron and silicon, the specific heat was found to be much too low even at room temperature. All this had been discovered in the 1870s by a former physics teacher of Einstein in Zurich, Heinrich Weber (one of the professors he dismissed as fossilized, who had refused to teach him Maxwell's equations!).

Einstein decided to seek a quantum explanation of these specific heat anomalies. According to the classical kinetic theory (discussed earlier in relation to Brownian movement in liquids and gases), the atoms in solids must absorb heat by vibrating more vigorously on their crystal lattices, in the same way that atoms in liquids and gases zip around with higher velocities at higher temperatures. But supposing the fixed atoms in

a solid oscillated not in a continuous manner but in a quantized way, such that they could increase their energy of vibration only in steps, not continuously? In other words, suppose the vibrations could take only discrete energy values. Einstein further assumed that the magnitude of the quantized energies could be calculated from Planck's simple quantum equation linking energy and frequency (in this case the frequency of atomic vibration), an equation as familiar to modern physicists as Newton's second law of motion:

$E = h\nu$

Energy = Planck's constant x *frequency.*

On this basis his calculation gave "a remarkably accurate account of the general behaviour of simple solids," as the solid-state physicist Philip Anderson explains on page 127 of this book.

Einstein's new model made predictions that matched his old teacher Weber's experimental data very well, especially considering that any model contains many unavoidable over-simplifications (in contrast, it is worth adding, to a theory like special relativity). At high temperatures his equation approximated the Dulong and Petit law, but as the temperature dropped it predicted a fall in the specific heat capacity like that actually observed, and that this should approach zero at very low temperatures (which is also observed). Einstein remarked in his 1907 paper: "If Planck's theory of radiation goes to the heart of the matter, then we must also expect to find contradictions between the present kinetic [i.e. classical] theory and experiment in other areas of the theory of heat—contradictions that can be resolved by following this new path."

Two years later, after further reflection and publication on the subject, at the Salzburg meeting of German scientists in 1909, instead of speaking on relativity as everyone had expected, Einstein's keynote address was on "The nature and constitution of radiation." It included this astonishingly prescient statement:

A picnic for Einstein in the woods near Oslo, 1920. With him is Victor Goldschmidt, the founder of modern geochemistry, who was interested in the application of Einstein's quantum theory to the solid state.

It cannot be denied that there exists a large group of radiation-related facts which show that light possesses certain fundamental properties which can much more easily be understood from the standpoint of Newtonian emission theory [i.e. light corpuscles/quanta] than from the standpoint of the wave theory... It is my opinion that the next phase in the development of theoretical physics will bring us a theory of light that can be interpreted as a kind of fusion of the wave and the [particle] theory... [The] wave structure and [the] quantum structure...are not to be considered incompatible.

The Salzburg speech was "one of the landmarks in the development of theoretical physics," said Wolfgang Pauli, another pioneer of quantum mechanics, on Einstein's seventieth birthday 40 years later. It was the first announcement of the disturbing idea of wave-particle duality.

Soon after this, when he moved to Prague in 1911, Einstein shifted his attention to general relativity for the next five years, though of course he continued to think about quantum theory. He had no role at all in the next important quantum development, which came from Bohr, who was working in Rutherford's laboratory on the structure of the atom.

In 1911, Rutherford had discovered the atomic nucleus and conceived the solar-system model of the atom, in which an electron was thought to orbit the nucleus like a planet around the Sun, held in place not by gravity but by an electrostatic force acting between the positively charged nucleus and the negatively charged electron. This was a familiar enough concept from the classical physics of the nineteenth century. But, appealing though this picture was (as any high-school physics student knows), it did not explain atomic spectra, which had been documented over many decades. That is, the model did not explain why atoms of different elements should emit and absorb electromagnetic radiation at specific frequencies/wavelengths, creating sharp bright and dark lines in the electromagnetic spectrum that are as characteristic of a given element as a fingerprint is of a human individual. (The colours of fireworks illustrate the visible emission spectra of different chemical elements and compounds, as does a sodium or neon street light.)

According to Bohr's proposal of 1913, the existence of spectra implied that the electrons in an atom could occupy only particular orbits, not continuously varying orbits. Then emission of light—a bright spectral line—would occur when an electron fell from one fixed orbit to another lower-energy orbit nearer the nucleus, thus decreasing its energy and emitting the balance of its initial energy as radiation of a characteristic frequency. Absorption of light—a dark spectral line—would entail the reverse of this process, with an electron jumping to a higher-energy orbit further away from the nucleus.

In order to explain these fixed orbits, Bohr, like Planck before him in 1900, was obliged to add a crucial postulate of his own: that the electron orbits of Rutherford's atom were quantized. The angular momentum (and hence the velocity and energy) of an orbiting electron could take only certain values, Bohr decided. These values were whole-number multiples

Einstein with Wolfgang Pauli, a founder of quantum mechanics.

of a constant based on Planck's constant *h*, which appears in Planck's equation connecting energy and frequency. On this basis Bohr successfully constructed a model of a one-electron hydrogen atom with orbital energy levels that accounted for the hydrogen spectrum that had been observed. In 1916 Einstein called this a "revelation."

There were two major weaknesses though. First, the model offered no convincing explanation for the stability of atoms. According to Maxwell's equations electrons, being accelerated charged objects, must radiate energy and quickly spiral into the nucleus. Bohr's quantum postulate forbade such an atomic collapse by simple fiat. Secondly, although the electron orbits themselves were quantized, the radiation emitted and absorbed by the electrons was observed in the form of (continuous) *waves*. Even if one accepted this discrete/continuous contradiction, how could an electron 'know' when it left one orbit at what frequency it should immediately begin radiating before it had 'arrived' at its next orbit? How did it know to which orbit it was going? What really occurred during the transition? "Such questions plagued the theory," writes a current physicist (Whitaker). "In modern quantum theory we learn, not so much how to answer them, as not to ask them." At this early stage of the quantum theory Bohr had no confidence in quantized *light* (he was one of the last physicists to take Einstein's photons seriously in the 1920s)—only in quantized electron orbits. In the end, Bohr's model of the atom blended classical and quantum physics imaginatively, even brilliantly (as a Nobel prize soon confirmed), but without fully satisfying anyone.

Einstein with a group of scientists in Belgium in 1932, making plans for the 1933 Solvay Congress. To his left is Niels Bohr, whom Einstein admired but radically disagreed with concerning quantum theory. The photograph was taken by Queen Elisabeth of Belgium, who became friendly with Einstein; they played music together and corresponded regularly from 1929 until Einstein's death.

Ernest Rutherford at the Cavendish Laboratory, Cambridge, where he was the director from 1919. The sign "TALK SOFTLY PLEASE" was a laboratory joke; Rutherford was famous for his booming voice, which could disturb both his researchers and their delicate equipment.

Einstein saw this, and saw he could connect Planck's radiation law with Bohr's atomic energy levels using his light quanta. In 1916–17, having completed general relativity, he published three papers using the concept. His new idea was that atoms, in addition to emitting light spontaneously, could be forced or *stimulated* by light to emit light, in the process moving from a high-energy state to a lower-energy state. The effect would be that *one* light quantum would stimulate an atom and *two* quanta would emerge, resulting in light amplification. The laser—which stands for 'light amplification by stimulated emission of radiation'—eventually grew from this idea. "Einstein was the first to recognize clearly, from basic thermodynamics, that if photons can be absorbed by atoms and lift them to higher energy states, then it is necessary that light can also force an atom to give up its energy and drop to a lower level. One photon hits the atom, and two come out," wrote Charles Townes, one of the inventors of the laser in the 1950s, to which we shall return in Chapter 7. Einstein himself told his old friend Besso in 1916: "With this, the light quanta are as good as certain." But in his published paper he noted that his theory of stimulated emission had a failing: that it "leaves the time and direction of the elementary process to 'chance.'" Here was a hint of his coming dislike of quantum mechanics because of its random, probability-based foundations.

However before we finally reach that debate, there was yet a further prescient proposal from Einstein arising from light quanta. In 1924, he received a paper from an unknown physicist in India, Satyendranath Bose, entitled "Planck's law and the light quantum hypothesis." It derived Planck's radiation law not from classical electrodynamics but by regarding radiation as a gas composed of light quanta and then applying a statistical approach based on the fact that large numbers of photons—unlike, say, electrons—are permitted by nature to occupy exactly the same quantum state. Bose asked Einstein for help in getting his article published. Einstein immediately read it, translated it into German and recommended it to a journal—and then himself wrote two papers inspired by it, in which he applied Bose's approach to *atoms*. The results were called Bose-Einstein statistics and gave rise to a concept of a class of elementary particles now known as bosons (as opposed to a second class known as fermions, including electrons, protons and neutrons, which obey Fermi-Dirac statistics). In 1925 Einstein predicted that bosons, given the right conditions at very low temperature, could condense into a new state of matter. In Chapter 7, we shall come across the extraordinary properties of liquid helium acting as a superfluid and of the Bose-Einstein condensates known as 'superatoms,' which were at last produced in the laboratory in 1995, fully 70 years after Einstein's prediction.

With Bose-Einstein statistics we come to the end of the 'old' quantum theory and the beginning of quantum mechanics. It was also the end of Einstein's

Z. f. Phys. (4 Blatt Ms.) Kommt an Herrn Prof. Dr. A. Einstein, Berlin W. 30 Haberlandstr. 5 3. Juli 1924 (1)

#840.

Planck's Gesetz und Lichtquanten-Hypothese.

Von Bose (Dacca-University, Indien)

(Eingegangen am 2. Juli 1924)

Planck's Formel für die Verteilung der Energie in der Strahlung des schwarzen Körpers bildet den Ausgangspunkt für die Quanten-Theorie, welche in den letzten zwanzig Jahren entwickelt worden ist und in allen Gebieten der Physik reiche Früchte getragen hat. Seit der Publikation im Jahre 1901 sind viele Arten der Ableitung dieses Gesetzes vorgeschlagen worden. Es ist anerkannt, dass die fundamentalen Voraussetzungen der Quanten-Theorie unvereinbar sind mit den Gesetzen der klassischen Elektrodynamik. Alle bisherigen Ableitungen machen Gebrauch von der Relation

$$\varrho_\nu \, d\nu = \frac{8\pi\nu^2 d\nu}{c^3} E,$$

d. h. von der Relation zwischen der Strahlungs-Dichte und der mittleren Energie eines Oszillators, und sie machen Annahmen über die Zahl der Freiheitsgrade des Äthers, wie sie in obige Gleichung eingeht (erster Faktor der rechten Seite). Dieser Faktor konnte jedoch nur aus der klassischen Theorie hergeleitet werden. Dies ist der unbefriedigende Punkt in allen Ableitungen, und es kann nicht wundernehmen, dass immer wieder Anstrengungen gemacht werden, eine Ableitung zu geben, die von diesem logischen Fehler frei ist.

Eine bemerkenswert elegante Ableitung ist von Einstein angegeben worden. Dieser hat den logischen Mangel aller bisherigen Ableitungen erkannt und versucht, die Formel unabhängig von der klassischen Theorie zu deduzieren. Von sehr einfachen Annahmen über den Energie-Austausch zwischen Molekülen und Strahlungsfeld ausgehend findet er die Relation

$$\varrho_\nu = \frac{\alpha_{mn}}{e^{\frac{\varepsilon_m - \varepsilon_n}{kT}} - 1}.$$

Indessen muss er, um diese Formel mit der Planck'schen in Übereinstimmung zu bringen, von Wiens Verschiebungsgesetz und Bohrs Korrespondenz-Prinzip Gebrauch machen. Wiens Gesetz ist auf die klassische Theorie gegründet, und das Korrespondenz-Prinzip nimmt an, dass die Quanten-Theorie mit der klassischen Theorie in gewissen Grenzfällen übereinstimme.

Einstein's translation into German of the then-unknown Indian physicist Satyendranath Bose's paper "Planck's law and the light quantum hypothesis," dated 3 July 1924. This paper led Einstein to develop Bose-Einstein statistics and to predict the Bose-Einstein condensate.

central role in quantum theory; after 1926 he became a critic of, rather than a contributor to, the theory.

His criticisms turned out to be fundamental and fascinating. "Were it not for Einstein's challenge, the development of quantum physics would have been much slower," admitted Bohr in 1962. He and Einstein were the most persistent protagonists in this debate for some ten years, but at various times Einstein challenged just about every pioneering contributor to quantum mechanics. His long correspondence with Max Born (published by Born years after Einstein's death)—where Einstein first made what is probably his single most famous comment ever, "God does not play dice"—reveals just how keenly Einstein opposed quantum mechanics from beginning to end. In 1948, he told Born sternly that "if one abandons the assumption that what exists in different parts of space has its own, independent, real existence, then I simply cannot see what it is that physics is meant to describe."

Satyendranath Bose.

Modern quantum theory abandons exactly this assumption. In answer to Einstein's well-known poser "Does the moon really exist only when you look at it?" a current physicist, David Peat, answers: "Einstein's moon really exists. It is linked to us through non-local correlations but does not depend upon us for its actual being in the world. On the other hand, what we call the moon's reality, or the electron's existence, depends to some extent upon the contexts we create in thought, theories, language and experiments." In quantum theory, an electron (or a photon) has no independent reality 'out there,' wholly independent of the human world. It can be both a wave and a particle, depending on how it is measured and observed.

We shall now attempt to summarize in just one paragraph the development of quantum mechanics, a very complex and mathematically sophisticated subject, before returning to Einstein's critique. In 1924 de Broglie proposed that all matter has a wave associated with it, and this was quickly confirmed for electrons by electron diffraction experiments. In 1926 Schrödinger, in his classic wave equation (which we shall not frighten the reader by printing)—aided importantly by Born—replaced the picture of an electron as a particle having a precise position and momentum as it orbits a nucleus with a *wave function* that predicted stationary waves of electron probability around the nucleus. Schrödinger's equation enables physicists to calculate not the location of an electron at any given moment but its probability of being at any particular point in space. Hence the atom is no longer at risk of collapse, as in Bohr's original model, because no electric charge is being accelerated; the electron becomes a probability wave in the Schrödinger/Born model. Then in 1927 Heisenberg, in his far-reaching uncertainty principle, proved that the position and momentum of any elementary particle such as an electron can never be measured simultaneously with unlimited accuracy. The more an experimenter tries to pin down the position in space, the greater will be the uncertainty in the momentum, and vice versa, because the very act of observing the particle (say by firing a photon at it) will inevitably disturb its position and momentum. The Heisenberg uncertainty principle states that the uncertainty in the position multiplied by the uncertainty in the momentum will always exceed a constant based on Planck's constant h. Other kinds of uncertainty principle may also be derived, such as one which relates the uncertainty in the energy of a particle to the time interval in which one measures the energy.

At the Solvay Congress in 1927 and again in 1930, and after, Einstein tried to counter some of these ideas and their profound implications for physical reality with his own 'thought' experiments. Two of these are particularly well known. Here is the first one, from 1930 (we shall come to the second, which dates from 1935, in Chapter 7).

Imagine, said Einstein, a box containing radiation. There is a hole in its side and a shutter to open and close the hole. The box is weighed. Then the shutter is opened for a short time T and one photon is allowed to escape. The box is weighed again. The loss in mass, which must equal the mass of the photon, can be converted to a loss in energy (using $E=mc^2$). And this mass or energy can in principle be determined as accurately as one wishes, hence the uncertainty in the energy of the photon may be zero. The uncertainty in the time for the escape of the photon is finite, just T. This means that the two uncertainties multiplied together may be zero—in contradiction of the time-energy uncertainty principle.

Erwin Schrödinger, one of the pioneers of quantum mechanics.

"It was quite a shock for Bohr," remembered one of the other physicists at the 1930 Solvay Congress, Léon Rosenfeld. "During the whole evening he was extremely unhappy, going from one to the other, and trying to persuade them that it couldn't be true, that it would be the end of physics if Einstein were right; but he couldn't produce any refutation." Einstein left the meeting, "a tall majestic figure, walking quietly, with a somewhat ironical smile, and Bohr trotting near him, very excited."

But by the following morning, Bohr had a riposte ready. And it depended on general relativity! He had carefully considered, as Einstein had not, how the measurements of the loss in mass of the box and of the time interval for the shutter's opening and closing might actually be made by an observer. He imagined hanging the box from a delicate spring balance and attaching the shutter to a clock inside the box. Then he realized that general relativity dictated that the clock must change its rate as it moved very slightly upwards with the escape of the photon—since the clock was being decelerated in a gravitational field. This must introduce an uncertainty into the time interval T. As a consequence, Bohr's calculations showed, the time-energy uncertainty principle would be obeyed after all.

On this occasion Einstein was vanquished, and it appears that after 1930 he accepted that quantum mechanics was internally consistent. In 1931, he nominated Schrödinger and Heisenberg to the Swedish Academy for a joint Nobel prize as the founders of "wave, or quantum, mechanics" (note the uncertainty in the name). "In my opinion, this theory contains without doubt a piece of the ultimate truth," he explained. In 1932 Heisenberg received the prize, and in 1933 so did Schrödinger.

Yet we know that Einstein was far from satisfied with quantum mechanics, as he made abundantly clear for the rest of his life. For now, let him have the last word:

> *The conviction prevails that the experimentally assured duality of nature (corpuscular and wave structure) can be realized only by...a weakening of the concept of reality. I think that such a far-reaching theoretical renunciation is not for the present justified by our actual knowledge, and that one should not desist from pursuing to the end the path of the relativistic field theory.*

This was written in 1952. During his last three decades, Einstein had relentlessly pursued the path he had mentioned in an increasingly solitary search for a theory more fundamental than the new quantum theory.

Paul Dirac (fourth from left), Werner Heisenberg and Erwin Schrödinger at Stockholm train station, 1933, probably taken at the time Dirac and Schrödinger jointly received the 1933 Nobel Prize. Heisenberg had received the award a year earlier.

Attendees at the sixth Solvay Congress, Brussels, 1930, where Einstein had his famous debate with Niels Bohr over quantum mechanics. Also pictured are Marie Curie, Paul Dirac, Enrico Fermi, Wolfgang Pauli, Léon Rosenfeld and Arnold Sommerfeld.

6. The Search for a Theory of Everything

"No fairer destiny could be allotted to any physical theory, than that it should of itself point out the way to the introduction of a more comprehensive theory, in which it lives on as a limiting case."

Einstein, *Relativity*, 1916

A decade or so after Einstein's death, Max Born, in *The Born-Einstein Letters*—still probably the best book for understanding Einstein's intellect and personality—mentioned his late friend and sparring partner's latter-day search for a 'unified' theory in careful terms, as follows:

> *He saw in the quantum mechanics of today a useful intermediate stage between the traditional classical physics and a still completely unknown 'physics of the future' based on general relativity, in which—and this he regarded as indispensable for philosophical reasons—the traditional concepts of physical reality and determinism come into their own again. Thus he regarded statistical quantum mechanics to be not wrong but 'incomplete.'*

More colourfully, Einstein apparently told a former student (the astronomer Fritz Zwicky) that the aim of his search was "to obtain a formula that will account in one breath for Newton's falling apple, the transmission of light and radio waves, the stars, and the composition of matter."

To quote Einstein directly, the key concept for a unified theory had to be the "field," which had proved so

Voyaging to a new world: Einstein on board the SS *Deutschland*, on his second visit to the USA, 1931.

Max Born and Werner Heisenberg, pioneers of quantum mechanics. Both men disagreed strongly with Einstein about quantum theory, but while Born remained a very close friend, Einstein disliked Heisenberg, especially for his closeness to Nazi Germany.

fruitful in Maxwell's equations and in general relativity. Instead of particles of matter, he imagined regions of very intense field—rather like knots in the even grain of a piece of wood. In 1938 he wrote:

Could we not reject the concept of matter and build a pure field physics? What impresses our senses as matter is really a great concentration of energy into a comparatively small space. We could regard matter as the regions in space where the field is extremely strong. In this way a new philosophical background could be created. Its final aim would be the explanation of all events in nature by structure laws valid always and everywhere. A stone thrown is, from this point of view, a changing field, where the states of greatest field intensity travel through space with the velocity of the stone. There would be no place, in our new physics, for both field and matter, field being the only reality. This new view is suggested by the great achievements of field physics, by our success in expressing the laws of electricity, magnetism, gravitation in the form of structure laws and finally by the equivalence of mass and energy. Our ultimate problem would be to modify our field laws in such a way that they would not break down for regions in which the energy is enormously concentrated.

From his correspondence with various physicists it is clear that Einstein was seriously thinking about a unification of gravity and electromagnetism in such a field theory as early as 1918, and maybe as far back as 1916 after the completion of his great work on general relativity. In 1923 he referred to it in his Nobel lecture and published his first paper on it, only to abandon his ideas. Two years later he made another stab at it in a paper entitled "Unified field theory of gravitation and electricity." For some weeks he was again convinced he had the solution, but then brusquely declared it to be "no good." Further published attempts appeared in 1929, 1931 and 1950, all of which Einstein also rejected. On his deathbed he continued to work at a new one.

With each successive attempt the theory became more purely mathematical and less based on the real world. He had started his scientific life by imagining himself chasing a light ray, and had invented general relativity by wondering how gravity would feel if he jumped off a rooftop, but now he progressively lost interest in such physical ideas. Perhaps his flawed thinking in the 1930 'thought' experiment on quantum mechanics with the photon and the box was suggestive, in that the flaw lay in his failure to consider the physical method of measurement; by idealizing the experiment too much, he overlooked a key element.

Once—in the days of special relativity, light quanta and (less so) general relativity—Einstein had regarded mathematics chiefly in terms of its usefulness in rationalizing his physical understanding, as mentioned earlier. Now mathematics became for him the driving force behind a unified theory. In 1933, speaking at Oxford University, he went so far as to claim that:

Our experience hitherto justifies us in believing that nature is the realization of the simplest conceivable mathematical ideas. I am convinced that we can discover by means of purely mathematical constructions the concepts and the laws connecting them with each other, which furnish the key to the understanding of natural phenomena. Experience may suggest the appropriate mathematical con-

cepts, but they most certainly cannot be deduced from it. Experience remains, of course, the sole criterion of the physical utility of a mathematical construction. But the creative principle resides in mathematics. In a certain sense, therefore, I hold it true that pure thought can grasp reality, as the ancients dreamed.

During the 1930s and after, Einstein seems to have lost interest in the fundamental advances that were being made in physics. The discovery of the positron, the first known 'anti-matter' particle, in 1932–33 made little impact on his work—which was ironic given that Paul Dirac had predicted its existence in 1928 by applying special relativity to the quantum mechanics of the electron (though Einstein did admire Dirac's mathematics). And the same happened with the discovery of the neutron in 1932 and the muon in 1936, which heralded the discovery of many other nuclear particles. The richness of the newly discovered subatomic world—each particle with its mass, spin, charge, quantum number and other features—did not emerge from Einstein's new equations. Despite the importance of his 1905 equation $E=mc^2$ in understanding nuclear fission (discovered in 1938), he showed no serious interest in the emerging new model of the nuclear forces.

Einstein's two main mathematical approaches to a unified theory are outlined in this book by Steven Weinberg, who played a key part in the 'electroweak' theory of the 1960s and 70s that unified electromagnetism with the weak interactions of the nucleus. Weinberg concludes that while one of Einstein's approaches lives on, in much-modified form, in today's string theory, of the other approach "no trace remains in current research." Einstein's three decades of endless calculation after 1925 have left little behind except manuscripts. Although physicists may honour his final

Einstein with Paul Oppenheim in Davos, Switzerland, where Einstein collapsed with a heart condition in 1928, confining him to bed for four months.

search for its sheer faith in the possibility of unification—as is evident in today's search for a 'theory of everything'—his specific ideas were first ignored and then forgotten, in signal contrast with his work on relativity and quantum theory. The young Robert Oppenheimer privately thought Einstein "completely cuckoo" when he met him in 1935, while Bohr told a physicist in the 1950s that Einstein had become an "alchemist" —perhaps alluding to Newton's other great passion besides physics. In the 1990s, Einstein's scientific biographer Abraham Pais remarked that Einstein's fame in science "would be undiminished, if not enhanced, had he gone fishing"—or perhaps more likely sailing—after 1925.

So why did Einstein stick with his fruitless search, we may well ask? He was very well aware of what others thought of him in private. For instance, he mocked himself to Besso in 1938 for being "highly esteemed as an ancient labelled museum piece and curiosity." And to Born he joked in 1952: "One feels as if one were an Ichthyosaurus, left behind by accident."

Part of the reason may have been the stubbornness of an ageing physicist past his intellectual prime. There was also a sense of duty to physics. Einstein told a physicist who expressed regret at his efforts that though he knew the chance of success was very slight he felt obliged to try. "He himself had established his name; his position was assured, and so he could afford to take the risk of failure. A young man with his way to make in the world could not afford to take a risk by which he might lose a great career." But it seems likely that the main reason was the one suggested above by Born. Not only had Einstein always been drawn to the deepest questions in physics, he was also philosophically convinced that reality was determined by laws, not by chance, and that these laws made physical reality independent of the human mind. God does not play dice—he was certain.

Perhaps the clearest expression of this conviction came in Einstein's meeting with Rabindranath Tagore in Germany in 1930 not long before the Solvay Congress. Tagore, though best known as a philosopher, poet and song composer (for which he won the Nobel prize) and as a spiritual leader and fighter for India's freedom beside Gandhi and Nehru, was interested in science too. But Tagore's philosophical position was quite opposed to Einstein's. Their conversation, soon published in the *New York Times*, shines a bright light on Einstein's position and throws it into sharp relief.

"There are two different conceptions about the nature of the universe—the world as a unity dependent on humanity, and the world as reality independent of the human factor," said Einstein. Tagore replied: "This world is a human world—the scientific view of it is also that of the scientific man. Therefore, the world apart from us does not exist; it is a relative world, depending for its reality upon our consciousness."

"Truth, then, or beauty, is not independent of man?" "No," said Tagore. "If there were no human beings any more, the Apollo Belvedere no longer would be beautiful?" asked Einstein. "No." "I agree with regard to this conception of beauty, but not with regard to truth," said Einstein. "Why not? Truth is realized through men." "I cannot prove that my conception is right, but that is my religion," said Einstein firmly.

Einstein with Rabindranath Tagore, Berlin, 1930. While Einstein did not share Tagore's philosophical views, he was in sympathy with his social and political outlook and they met several times; they also shared a strong interest in music. Einstein's second wife Elsa and his step daughter Margot are present in the photograph on the left, as is Tagore's daughter-in-law (far right), and two Indian travelling companions, Prasanta and Rani Mahalanobis.

Niels Bohr and Einstein, taken by Paul Ehrenfest in about 1927.

Then he became concrete. "The mind acknowledges realities outside of it, independent of it. For instance, nobody may be in this house, yet that table remains where it is." "Yes," said Tagore, "it remains outside the individual mind, but not the universal mind. The table is that which is perceptible by some kind of consciousness we possess."

"If nobody were in the house the table would exist all the same, but this is already illegitimate from your point of view, because we cannot explain what it means, that the table is there, independently of us... We attribute to truth a superhuman objectivity," said Einstein.

"In any case, if there be any truth absolutely unrelated to humanity, then for us it is absolutely non-existing," replied Tagore.

"Then I am more religious than you are!" exclaimed Einstein.

Bohr's view of reality—and the views of some other quantum physicists—had more in common here with Tagore's view than with Einstein's. For quantum theory maintains, like Tagore, that reality is dependent on the observer. In science, said Einstein, "we ought to be concerned solely with what nature does." Bohr, however, insisted that it was "wrong to think that the task of physics is to find out how nature *is*. Physics concerns what we can say about nature."

Soon after his conversation with Tagore, in a seventieth birthday message for him, Einstein wrote: "Man defends himself from being regarded as an impotent object in the course of the Universe. But should the lawfuless of events, such as unveils itself more or less clearly in inorganic nature, cease to function in front of the activities in our brain?" From an early age Einstein had believed human free will to be an illusion. He had a gut belief in the existence of a supreme law-giver—called God, if you will (as Einstein often did). Somehow this belief in determinism coexisted with his extreme individualism and exceptionally strong ethical values. "I have never been able to understand Einstein in this matter," wrote Born. In the second half of this book, on Einstein the man, we shall try to make some sense of the paradox. But first let us see how his ideas have influenced physics in the half-century since he died.

Einstein in Vienna, 1921.

Einstein's Search for Unification

Steven Weinberg

Some time in the 1970s I received a letter from a San Francisco dealer in rare books and manuscripts. Enclosed was a photocopy of a scientific article by Einstein, hand-written in German. The dealer asked me to read the article, and give him advice about its importance. I am no Einstein scholar, and I can read German only with great difficulty, but a quick look at the article showed that it was another one of Einstein's many attempts in the 1930s and 1940s to find a unified theory of the electromagnetic and gravitational fields. I wrote to the book dealer and told him that, while anything in Einstein's own hand must be of value, this particular paper was of no great importance in the history of science.

Albert Einstein was one of the greatest scientists of all time, a peer of Galileo, Newton, and Darwin, and certainly the leading physicist of the twentieth century. It is not surprising that, after his scientific triumphs in the years from 1905 to 1925, he would turn his attention to the search for a unified theory of electromagnetism and gravitation. The greatest advances in the history of physics have been marked by the discovery of theories that gave a unified explanation of phenomena that had previously seemed unrelated. In the seventeenth century, Isaac Newton had unified celestial and terrestrial physics, showing that the same force of gravitation that causes apples to fall to the ground also holds the Moon in its orbit around the Earth and the Earth and planets in their orbits around the Sun. In the nineteenth century, James Clerk Maxwell had unified the phenomena of electricity and magnetism, by realizing that just as oscillating magnetic fields produce electric forces so also oscillating electric fields produce magnetic forces, and had discovered that light was nothing but a self-sustaining oscillation of both electric and magnetic fields. In 1915, Einstein himself, in his general theory of relativity, had shown that gravitation was but an aspect of the geometry of space and time. After that brilliant success, the obvious next problem was to find a theory that gave a unified account of gravitation and electromagnetism. Tragically, Einstein spent almost all of the last 30 years of his life pursuing this aim, not only without success, but without leaving any significant impact on the work of other physicists.

In this work Einstein followed two main lines of attack.

First, Einstein was attracted by an idea in a 1921 paper of the mathematician Theodor Kaluza, that electromagnetism could be understood as an aspect of gravitation in *five* rather than the usual four dimensions of space and time. It was clear from Einstein's development of general relativity that gravitation in any number of dimensions should be described by a matrix called the metric, a symmetric square array g_{MN} of numbers depending on position in space and

time. (M labels the rows and N labels the columns of the array; 'symmetric' means that $g_{MN} = g_{NM}$.) In five dimensions M and N run over the labels 1, 2 and 3 that are conventionally used to distinguish the three directions of ordinary space, plus a label 0 used to distinguish the dimension of time, plus a label 5 for the fifth dimension. Kaluza proposed that the part of g_{MN} in which both M and N run only over 1, 2, 3 and 0 represents the gravitational field observed in four space-time dimensions; that the quantities g_{51}, g_{52}, g_{53} and g_{50} form the 'vector potential' from which the electromagnetic field can be derived in conventional theories of electrodynamics; and that g_{55} is a field representing some sort of matter. Indeed, with this interpretation of the fields g_{MN}, if one arbitrarily assumes that these fields are independent of position in the fifth dimension, then the obvious extension of the field equations of general relativity to five dimensions yields the equations of general relativity for the gravitational field and the Maxwell equations for the electromagnetic field, both in four dimensions.

Not only did this theory seem to provide a unification of gravitation and electromagnetism; a subsequent extension of the Kaluza theory due to the physicist Oskar Klein in 1928 seemed to offer hope of solving a problem that

Einstein working on a calculation with his assistant Peter Bergmann at Princeton University, 1940.

had bedevilled Einstein since the advent of general relativity. In Einstein's 1915 formulation of general relativity, gravitation appears as a natural consequence of the geometry of space and time, and (aside from possible effects that would vanish at sufficiently large distances) the gravitational field equations are nearly unique.* On the other hand, matter is put into the theory 'by hand'; there is no *a priori* way of judging what sorts of matter exist, or how they contribute to the source of the gravitational field. What Klein suggested is that the fifth dimension of Kaluza is not just a formal device, but a real dimension of space, which is curled up so that it is not normally observed, just as a soda straw may appear one-dimensional when observed casually, but on closer inspection is found to be a two-dimensional sheet, with one dimension curled up. According to this idea, the fields g_{MN} are no longer arbitrarily constrained to be independent of position in the fifth, curled-up, dimension. Rather, they are superpositions of terms that *are* independent of the fifth dimension, which as Kaluza had found represent the four-dimensional gravitational and electromagnetic fields, plus new terms that oscillate in the curled-up fifth dimension, like sound waves in an organ pipe, with wavelengths that fit once, twice, or more times into the circumference of the curled-up dimension. In four dimensions these oscillating terms appear like an infinite variety of massive particles, with electric charges proportional to their masses. So not only gravitation and electromagnetism but also charged massive particles arose out of a purely gravitational theory in five dimensions.

The trouble was that the particles predicted by the Kaluza-Klein theory could not be electrons or protons or any other known type of elementary particle. Even the lightest of these new particles were predicted to be much too heavy, by about 19 orders of magnitude. They were so heavy that the gravitational attraction between pairs of these particles would be as strong as their electric attraction or repulsion, which is certainly not the case in ordinary atoms. This setback seems to have caused Einstein to lose interest in theories with extra dimensions, and his efforts in the 1940s turned in a different direction.

In his second main approach to unification, instead of increasing the number of space-time dimensions, Einstein considered the possibility that the metric g_{MN} may not be constrained to be symmetric. (In some of his work, Einstein considered the equivalent possibility that the metric has a mathematical property known as hermiticity.) His motivation was an elementary bit of counting. A symmetric 4 x 4 array has ten independent entries: g_{11}, g_{22}, g_{33}, g_{00}, $g_{12}=g_{21}$, $g_{23}=g_{32}$, $g_{31}=g_{13}$, $g_{10}=g_{01}$, $g_{20}=g_{02}$ and $g_{30}=g_{03}$. But without the constraint of symmetry, a 4 x 4 array has

*I say "nearly unique," because in their original 1915 form Einstein's field equations omitted one possible term, known as the cosmological constant, that would affect phenomena at great distances. In 1917 Einstein introduced the cosmological constant, to account for what seemed to be an absence of any large-scale motion of matter in the universe. A few years later, after the discovery of the expansion of the universe, Einstein came to regret the introduction of the cosmological constant, but recent studies of the expansion of the universe strongly suggest that it does appear in the field equations after all.

The German mathematician Theodor Kaluza, whose work inspired one of Einstein's unsuccessful attempts at a unified theory.

$4^2 = 16$ independent entries. Einstein supposed that the $16 - 10 = 6$ extra fields in a general non-symmetric metric might represent electromagnetism, which is described by the three components of the electric field and the three components of the magnetic field. It was this hope that preoccupied Einstein in the last decades of his life.

The trouble with this idea is that the 10 components of the symmetric part of a general metric and its additional 6 components have nothing to do with each other. Just putting them together in a 4×4 array says nothing about how their physical properties are related. This is very different from Maxwell's unification of electric and magnetic fields. For one thing, what looks like a purely electric or magnetic field to an observer at rest will look like a combination of electric and magnetic fields to a moving observer, while in Einstein's new theory what looked to one observer like a purely gravitational field would look like a purely gravitational field to all observers. Einstein of course understood this very well, and he kept searching for some physical principle that would tie all 16 components of the metric together in a natural way, but he never succeeded.

As it has happened, over the half-century since his death, Einstein's dream of unification has been partly realized, but in a way very different from what Einstein expected. The theory of electromagnetism is now understood to be part of a larger *electroweak* theory, which describes not just electromagnetism but also certain weak nuclear forces. These forces are responsible for the radioactive processes in atomic nuclei in which neutrons turn into protons, or vice versa. The weak nuclear forces have very short range—the weak force between two nuclear particles drops off steeply at a distance beyond about a quadrillionth of a centimetre; in contrast, electromagnetism like gravitation is a long-range force—the force of attraction between two charged particles falls smoothly as the inverse square of the distance between them, with no steep drop anywhere. Nevertheless, despite this obvious difference between the electromagnetic and weak nuclear forces, they enter in the same way in the modern electroweak theory, and the difference between them is attributed to properties of the space in which we live, rather than to the theory itself.

Einstein of course knew about radioactivity. It was discovered in 1897, and radioactive salts provided him a decade later with a vivid example of the relation $E = mc^2$ between mass and energy. But as far as I know, Einstein never concerned himself with understanding the weak nuclear forces that make

Einstein receives a citation as a "modern pioneer of science" at the Copernican Quadricentennial in Carnegie Hall, New York, 1943. Other recipients at the ceremony included Walt Disney, Henry Ford and Orville Wright. Einstein wore the academic regalia of the Sorbonne to show sympathy with France, then under Nazi occupation.

radioactivity possible. Indeed, in his later years Einstein showed no interest in any contemporary work on nuclear and particle physics. Perhaps this was because this work was neither based on nor incorporated general relativity. Einstein remarked in 1950 that "all attempts to obtain a deeper knowledge of the foundations of physics seem doomed to me unless the basic concepts are in accordance with general relativity from the beginning." Also, the work of other physicists on nuclear and particle physics was grounded in quantum mechanics, the new probabilistic approach to theoretical physics developed in the 1920s. Einstein regarded quantum mechanics as a renunciation of the traditional aim of physics, to come to a complete understanding of physical reality. Indeed, one of Einstein's hopes for a unified theory was that it would provide an alternative non-quantum mechanical explanation of the atomic phenomena that had already been successfully accounted for by quantum mechanics.

Oskar Klein (standing far right) at the "Bohr Festspiels", Göttingen. Niels Bohr is standing second from left; seated is Max Born.

Particle physicists in the 1970s also developed a highly successful theory of one other sort of force, the strong nuclear force that holds quarks together inside the neutron and proton and holds neutrons and protons together inside atomic nuclei. This theory, known as quantum chromodynamics, is mathematically similar to the electroweak theory, so it has not been difficult to imagine the unification of both theories in a so-called 'grand unified' theory of the electromagnetic, weak and strong nuclear forces.

However it has been much more difficult to bring gravitation into this general theoretical framework. The superficial similarity between gravitation and electromagnetism, both producing forces that fall off with the inverse square of the distance, has proved illusory. The one hope for a unification of gravitation with the other forces of nature now lies in the direction of a string theory, a theory that supposes that the fundamental constituents of nature are neither particles nor fields, but strings, one-dimensional entities that are much too small to be seen by us as anything but point particles, but whose various modes of vibration account for the variety of the particle types we observe.

Ironically, these string theories find their most natural formulation in a space-time of ten rather than four dimensions, thus reviving the Kaluza-Klein idea that had so attracted Einstein in the 1930s, though with six extra dimensions rather than one. But of Einstein's other approach to unification, the extension of general relativity to a non-symmetric metric, no trace remains in current research. The idea of extra dimensions, especially in the form advanced by Klein, was highly speculative, and may have seemed at first like a purely mathematical game, but right or wrong it had real physical content. In contrast, the idea of a non-symmetric metric was purely mathematical, and led nowhere. In developing the general theory of relativity from 1905 to 1915, Einstein had been guided by an existing mathematical formalism, the Riemann theory of curved space, and perhaps he had acquired too great a respect for the power of pure mathematics to inspire physical theory. The oracle of mathematics that had served Einstein so well when he was young betrayed him in his later years.

7. Physics Since Einstein

"Science is not and never will be a closed book. Every important advance brings new questions. Every development reveals, in the long run, new and deeper difficulties."

Einstein, *The Evolution of Physics*, 1938

Relativity theory and quantum theory are generally recognized as the "two greatest revolutions in physical science since Galileo and Newton"—to quote Philip Anderson's article on Einstein's scientific legacy in this book. Einstein was central in developing both of these theories, so his work has, in a sense, influenced almost every aspect of physics in the half-century since his death—whether it concerns the structure and origin of the universe or the fission of the atom nucleus, the high-energy acceleration of subatomic particles or the photoelectric effect, the black hole or the microchip. This is perhaps not so surprising, if still unparalleled by any other physicist, when we recall that relativity theory unifies such fundamentals as space, time, gravity and the speed of light, while quantum theory unifies the constitution of matter and the transmission of radiation and energy.

Come down to specific influences, though, and Einstein's record is startling. Just consider how far advanced he was in how many disparate areas. In 1905, in his follow-up to special relativity deriving the equation $E=mc^2$, Einstein perceived and in essence predicted nuclear energy 33 years before the discovery of nuclear fission (in 1938). In 1916, in his work on

High-energy particle acceleration is a field of physics strongly influenced by relativity. Here a magnet coil arrives at the site of CERN, the European particle physics laboratory, near Geneva, Switzerland, in 1955, the year of Einstein's death.

general relativity, he predicted the gravitational red shift, which was not confirmed for almost half a century (in the early 1960s). In 1917, in his paper on the stimulated emission of light, he conceived the possibility of a phenomenon not demonstrated until 37 years later, which then led to the laser (in 1958). In 1925, in his two papers on Bose-Einstein statistics, he envisaged a new state of matter, the Bose-Einstein condensate, which took 70 years to be observed (in 1995) and today offers promise of its first application—as has happened many times with the laser—as a so-called 'atom interferometer' for the measurement of small changes in the strength and direction of gravity. In 1935, in his 'EPR' paper written with Boris Podolsky and Nathan Rosen (see page 119), he introduced the controversial concept of what became known as quantum entanglement, a subject for experimental research three decades later which has potential applications in quantum computing, quantum cryptography and even, some claim, teleportation.

And this leaves aside Einstein's 1918 prediction of gravitational waves. Despite decades of intensive searching, no waves have been detected. The problem does not seem to lie with the theory but with the minuteness of the ripples of gravitational energy to be detected. The ground-based Laser Interferometer Gravitational-Wave Observatory (LIGO) physically resembles the Michelson-Morley experiment of the nineteenth century—except that ordinary light beams have been replaced by lasers. The sensitivity depends on the length of the laser beam. The best hope of detecting gravitational waves therefore lies with a satellite-based system in which the laser beams are much longer than LIGO's. The plan is to build a Laser Interferometer Space Antenna (LISA), having laser arms forming an equilateral triangle with sides of about three million miles, orbiting the Sun at about the same distance as the Earth. If LISA gets launched around 2010, as intended, then an unequivocal observation of 'gravity's shadow' is likely before the centenary of Einstein's prediction.

A radio telescope on the volcano Mauna Kea in Hawaii, part of the Very Long Baseline Array (VLBA) which has been used to test some of the astronomical predictions of general relativity. The VLBA's ten identical antennae, located at stations from Hawaii to the Virgin Islands, simulate a giant antenna by using interferometry techniques to combine the data obtained by each individual telescope.

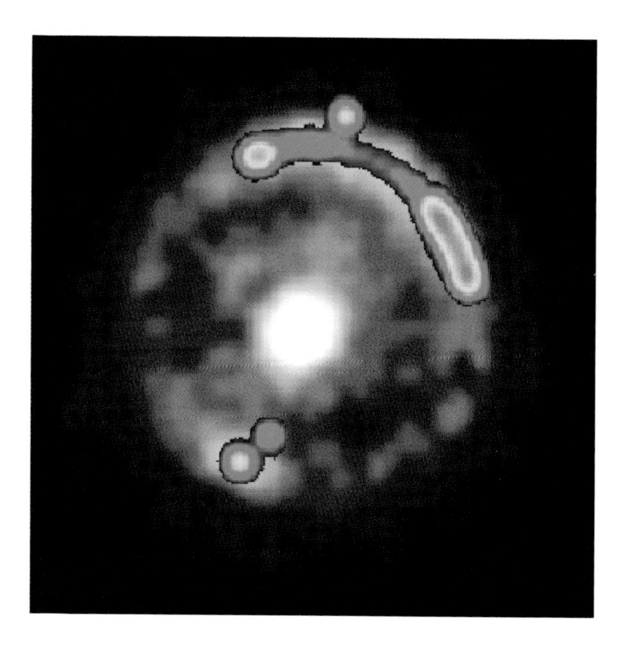

A combined radio and infrared image of an 'Einstein ring' (formed as a result of a gravitational lens) in the system B1938+666. Here, a huge galaxy (white) is positioned between Earth and another galaxy. Infrared (orange) and radio frequency (blue, green) radiation from the distant galaxy is bent around the nearer galaxy by its gravitational pull, forming a ring.

A recent productive experiment on gravity exemplifies the reach of Einstein's ideas. In 2002, two scientists working together—Ed Fomalont, a radio-astronomer, and Sergei Kopeikin, a theoretical physicist—used another idea first mooted by Einstein for a purpose not originally suggested by him: the measurement of the speed of gravity. Newton of course assumed that gravity was instantaneous, in other words its speed was infinite, while Einstein assumed that it moved at the speed of light. Kopeikin realized that it was possible for Einstein's equations of general relativity "to be reformulated in a way that made gravity analogous to electromagnetic radiation." With light, one can work out its speed by measuring its electric field and its magnetic field. With gravity, too, Kopeikin reckoned, it should be possible to work out the speed if one could measure in detail the gravitational field of a massive moving body.

Inside the tunnel being made for the Super Proton Synchrotron (SPS) at CERN, 1974. The SPS was the first of the large accelerators at CERN, an underground ring with a 7-kilometre circumference, which accelerated particles to high energies, smashing them into targets or colliding two particle beams together.

The idea that interested these two scientists had been published by Einstein in a short paper in the journal *Science* in 1936, entitled "Lens-like action of a star by deviation of light in the gravitational field." Back in 1915, while working on general relativity, he had proposed that light from a star could be bent by the gravitational field of the Sun (as confirmed in the world-famous eclipse of 1919). Once again his 1936 idea was that the gravity of a massive body—this time not the Sun—could act like a lens on light coming from a distant star if there was an exact alignment between the star, the massive body and the observer on Earth. Then two images of the distant star should be visible through telescopes: one in its normal position and the other in its deflected position. Einstein was doubtful whether a precise enough alignment could actually occur for deflection to be observed and most astronomers agreed with him in the 1930s. But in the 1960s came the discovery of quasars: 'quasi-stellar' objects in remote regions of the universe that are intense emitters of radio waves (some think they are powered by super-massive black holes). In 1979 and after, a number of quasars with identical twin images were identified and it became clear to astronomers that these were the result of a 'gravitational lens.' The radio waves from the distant quasars were being bent by closer galaxies and clusters of galaxies acting like lenses on the electromagnetic radiation as it passed through space towards the Earth.

Fomalont and Kopeikin decided to arrange a kind of gravitational lens trick. They planned to line up a bright radio quasar, the giant planet Jupiter and Earth's most powerful intercontinental array of radio telescopes. The hope was to detect the distortion by Jupiter's gravitational field of the radio waves from the quasar. They knew that Jupiter would be a good choice as a 'cosmic lens' for measuring gravity in precise detail because its mass and orbital velocity were known with tremendous accuracy as a result of the fly-bys of the *Pioneer*, *Voyager* and *Galileo* spacecraft.

In 2000, by comparing the orbit of Jupiter for the next 30 years with catalogues of suitable quasars, Kopeikin discovered that one particular quasar known as J0842+1835 would align with Jupiter and Earth on 8 September 2002 at 16.30GMT. When the time came, the predicted gravitational lens action was observed, despite the malfunctioning of one of the radio telescopes in the US Virgin Islands and the loss of 15 per cent of the data due to bad weather. The rest of the data, when plugged into Kopeikin's reworked equations of general relativity, gave a figure for the speed of gravity of 1.06 times the speed of light. Since their calculated measurement error was plus or minus 0.21, Fomalont and Kopeikin concluded that gravity probably does move at the same speed as light. Certainly it cannot move instantaneously because were it to do so they would have observed a different—admittedly a minutely different—split in the double image of the quasar.

Although general relativity is now a key part of the cosmology, and although special relativity is essential in many parts of physics, especially high-energy particle physics, it needs to be said that relativity still has its critics. Einstein made a significant number of mistakes in his scientific career, as he was usually willing to admit, and this fact is well known among physicists, which presumably encourages the critics of relativity. Stephen Hawking gets regular letters telling him Einstein is wrong, as he mentions in this book. When the physicist John Rigden was editor of the *American Journal of Physics* from 1978 to 1988, he received "scores of manuscripts from authors who attacked relativity theory and purported to reveal various errors that Einstein had made." Their motives varied:

> *Many of these manuscripts were driven as much by emotion as by intellectual considerations. Many find the consequences of the special theory too abstract and too contrary to common sense. Others accepted the theory, but rejected its implications. Still others could not accept a theory that was, in essence, the product of pure thought rather than of hard experimental facts. Some simply could not give up the ether. Reactions varied from one national setting and intellectual climate to another.*

Fig. 6: On 8 September 2002, Jupiter passed between Earth and the distant quasar J0842+1835. Using an array of telescopes on Earth including the Very Long Baseline Array, Sergei Kopeikin and Ed Fomalont measured the resulting gravitational lens effect and thereby estimated the speed of gravity.

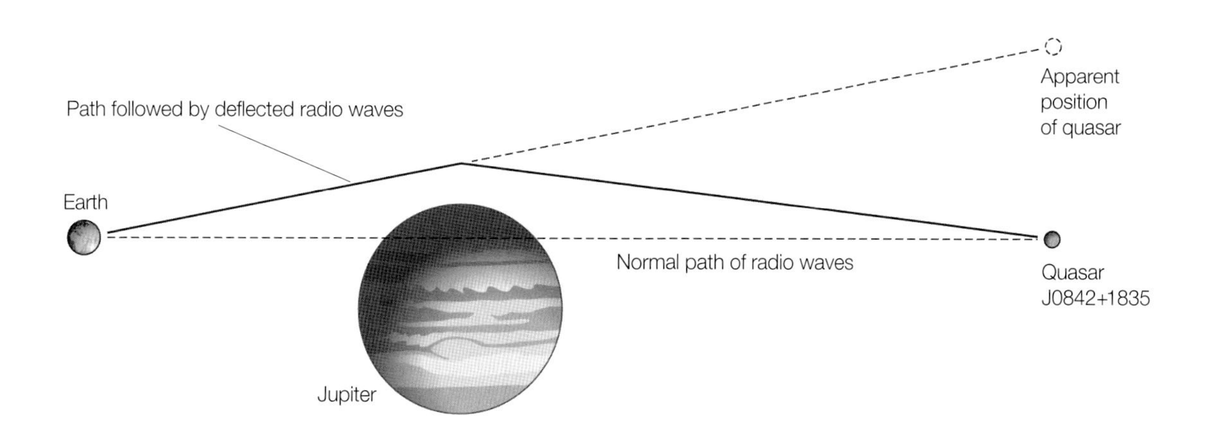

Einstein, aboard the SS *Belgenland*, 1930–31.

Opposite: Einstein with Edwin Hubble and others at Mount Wilson Observatory, California, 1931, when he first saw evidence of the expanding universe and at once abandoned the 'cosmological constant.'

The vast majority of these criticisms of relativity are ill founded. But as the physicist João Magueijo reminds us in this book, "as long as we have a good reason, we should be ready to give up any scientific principle, no matter how sacred." Magueijo is one of a small number of physicists who think that in order to quantize gravity, and in order to explain how the large-scale features of the universe grew out of the big bang, the speed of light will have to be allowed to vary, at least in the very early universe. He feels encouraged by the fact that Einstein himself published more than one paper with a varying speed of light. Though Einstein soon realized his particular argument was wrong, these papers show that he was not dogmatic about the constancy of the speed of light.

By far the best known of Einstein's mistakes—because he deliberately drew public attention to it—is his 1917 addition of the cosmological constant to his 1915 field equations of general relativity. He added this 'fudge factor' only in order to make his proposed model of the universe static and therefore eternal, which is how he thought the universe should be. Since the constant was "necessary only for the purpose of making a quasi-static distribution of matter," he naturally regarded it as an ugly intrusion into the elegance of his original equations. Then in the 1920s Edwin Hubble's telescopic observations of galaxies proved that the universe is expanding, not static. In 1931, on a visit to the Mount Wilson Observatory in California, Einstein saw the photographs of galaxies taken by Hubble and his assistant Milton Humason. The evidence convinced him and immediately he told a group of journalists that he would abandon the static model of the universe in favour of an expanding one. The cosmological constant could now be set at zero. Though Einstein much regretted having included it in the first place, he welcomed the increase in simplicity for his field equations.

"However the possibility of a cosmological constant did not go away so easily," Steven Weinberg wrote in 1993.

> *Einstein in 1915 operated under the assumption that the field equations should be chosen to be as simple as possible. The experience of the past three quarters of a century has taught us to dis-*

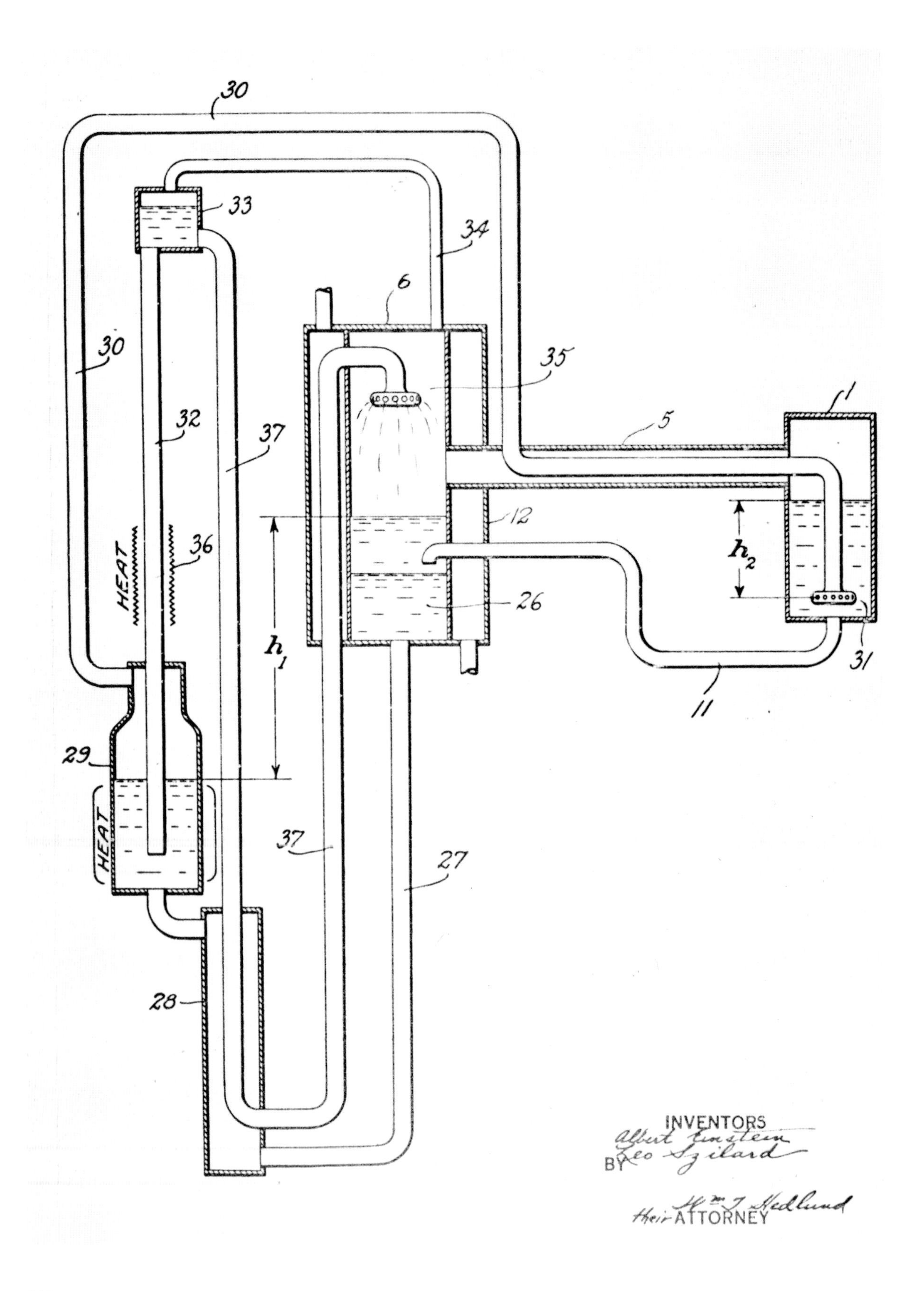
30
33
34
6
30
35
1
32
5
37
12
36
HEAT
h2
26
h1
31
11
29
HEAT
37
27
28
INVENTORS
Albert Einstein
Leo Szilard
BY
Wm. J. Hedlund
their ATTORNEY

trust such assumptions; we generally find that any complication in our theories that is not forbidden by some symmetry or other fundamental principle actually occurs. It is thus not enough to say that a cosmological constant is an unnecessary complication. Simplicity, like everything else, must be explained.

Weinberg's caution was justified. Recent observations, as both he and Hawking mention in this book, suggest that the cosmological constant is not zero. This appears to be connected with the fact that the expansion of the universe is accelerating.

Einstein now favoured an expanding universe, which necessarily entailed a beginning to the universe. Yet it would not be true to say that he was interested in applying general relativity to cosmology. As Hawking also mentions, Einstein did not apparently take seriously the 'big bang' origin of the universe (as it became known in the early 1950s), and did not welcome the idea that time might have a beginning. Perhaps more surprisingly he rejected black holes, which would become a *tour de force* for general relativity in the work of Roger Penrose, Hawking and others in the 1960s and 70s. In his preface, Freeman Dyson expresses bafflement that Einstein was totally indifferent to an early (1939) relativistic calculation by Robert Oppenheimer and Hartland Snyder which supported the existence of black holes: "how could he have been blind to one of the greatest triumphs of his own theory?" Obviously Einstein did not have the benefit of the compelling astronomical observations of the 1970s and after, which strongly suggest that black holes are common. However a lack of observational evidence had never stopped Einstein before when he was enamoured of a theory. The theoretical physicist Kip Thorne suggests a different possible answer: "It was clear that

A design by Einstein and Leo Szilard for a refrigerator pump, around 1927. By avoiding the use of moving parts, the inventors made the pump leak-proof. They sold the first two designs to the Electrolux Company and a later design to AEG, the German division of General Electric; but in 1932, after five designs in all, development was abandoned, partly because of the invention of Freon, a safer coolant. Later, the Einstein-Szilard model was used for the circulation of liquid sodium coolant in nuclear reactors.

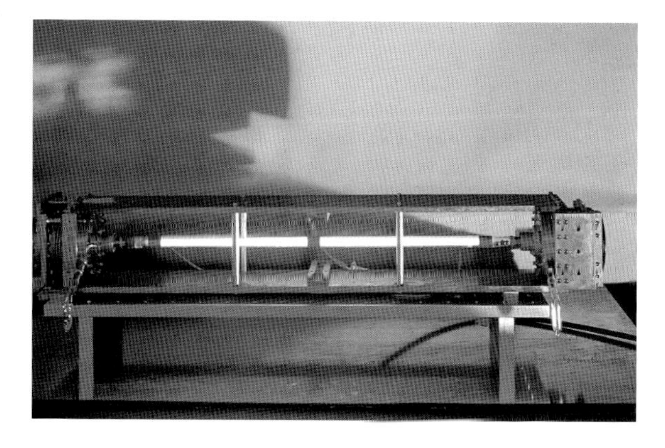

An early laser, built in the USA by Theodore Maiman, 1960. Although Einstein did not invent the laser himself—others, including Charles Townes (below), take the credit—his work on light quanta and stimulated emission were crucial to its development, as Townes was happy to acknowledge.

anything that falls into a black hole can never get back out and cannot send light or anything else out, and this was enough to convince Einstein and most other physicists of his day"—including Arthur Eddington—"that black holes are outrageously bizarre objects which surely should not exist in the real universe. Somehow, the laws of physics must protect the universe from such beasts."

But enough of Einstein's role in cosmology—let us return to Earth. A remarkable number of important technologies have sprung from the theoretical ideas of the former patent clerk. (His patents were rather less influential, though his 1920s gyrocompass design was used by the German navy and other national navies and his leak-proof refrigerator pump designed with Leo Szilard, also in the 1920s, though never commercially manufactured, was later used in the cooling system of fast-breeder nuclear reactors.) To mention only the photoelectric effect that Einstein explained in his first published paper in 1905, its applications now include the ubiquitous solar cell, the highly sensitive photomultipliers used in astronomical detection and the devices that switch on street lights at dusk, control the toner density in photocopying machines, expose photographic film to light correctly and detect the colour change in a Breathalyser that appears after alcohol reacts with a test gas. "A hundred years later technologists are still finding new ways to harvest novel inventions from Einstein's theories," noted an Einstein special issue of *Scientific American*.

His two 1905 papers on molecular dimensions and movement have spawned practical applications also. Indeed the first one, the doctoral thesis ("A new determination of molecular dimensions"), which deals with the bulk rheological properties of particle suspensions—how they flow and deform—has been very widely cited in situations ranging from the motions of sand particles in cement mixes in the construction industry to the motions of casein (protein) micelles in cow's milk in the dairy industry and those of aerosol particles in clouds in ecology. His second paper, the one on the Brownian movement, which showed mathematically what is happening to small particles when they are seen to move randomly in liquids—a much more important paper than the doctoral thesis for its demonstration of the existence of atoms and molecules—has not had the same practical impact. However, recently it has led to research on improved methods of sieving and sorting liquids containing pollutants—for example, soot, viruses, cell fragments and large fragments of DNA—with obvious relevance to blood and water purification. So-called 'Brownian ratchets' have been designed. Their basic principle is that Brownian movement over a given time displaces small particles further than large particles, as calculated by Einstein. When a liquid is passed through microscopic channels containing obstacles, Brownian movement causes the particles to bump into the obstacles and become differentially separated; the obstacles act like the teeth of a ratchet gear that can move only in one direction.

The laser emerged from Einstein's 1905 light quantum hypothesis that explained the photoelectric effect. But it required, besides clever and tricky engineering, two further theoretical concepts: stimulated emission, which he put forward in 1917, and *coherence* of the emitted light, which he did not. Einstein predicted that one photon could stimulate an atom in a high-energy state to emit two photons of the same energy, but he did not indicate that they would be, so to speak, identical copies. Not only would they have the same frequency (since their energy was the same), they would also be in step with each other, in other words the wave peaks and troughs would be similarly spaced, with a fixed phase relationship. A coherent beam of light consists of waves that are in phase; in an incoherent beam—that is, practically all light (sunlight, lamplight and so on)—the peaks and troughs of the waves are not aligned in the same way. It is the coherence and the single frequency of light in stimulated emission that gives a laser beam its power. The fact that Einstein did not consider coherence may well have prevented him from inventing the laser principle himself. Charles Townes, one of its inventors in the 1950s, thought that technologically speaking the laser could have happened in the 1920s. Ironically, when it was at last invented, it was not thought to be potentially useful. Four decades later lasers are used very widely outside the laboratory, for instance in eye surgery, cutting tools, fibre optics and DVD players.

Lasers were required to produce the first Bose-Einstein condensate, though its strange properties had been surmised from the superfluidity of liquid helium, which can flow with almost no viscosity below a temperature of about two degrees above absolute zero. A beaker of liquid helium forms drops on the outside of the beaker at the bottom because the helium 'creeps' up the walls of the beaker, flows down the outside of the beaker and covers its entire surface, inside and out, with a thin film of liquid. In 1938, Fritz London suggested that this weird behaviour might be understood in terms of a Bose-Einstein condensation. However, true condensation required a still lower temperature, for which individual atoms (of rubidium not helium) had to be cooled by lasers. The basis of this surprising technique goes like this, the physicists Tony Hey and Patrick Walters explain:

Imagine an atom in the gas moving with exactly the right speed to absorb a photon from an approaching laser beam. When the atom absorbs the photon it will slow slightly from the impact. Of course the photon will ultimately be re-emitted but in a random direction. Since the laser beam contains many photons this process can be repeated many times. The overall effect is rather like the atom moving into a hail of bullets. The net effect is to slow the motion of the atom in the direction of the laser beam and add a small random motion in other directions.

By using six lasers and also magnetic fields (see Fig. 7), atoms can be slowed and trapped together in a tiny space. The slower they move, the lower their temperature. Eventually, at 20 thousand millionths of a degree above absolute zero, about 2000 rubidium atoms were induced to form a Bose-Einstein condensate in which they could be considered to be coherent (in phase) like a single atom, or 'superatom.' This was achieved in 1995. In 1997 and 2001, two Nobel prizes were awarded to the six physicists chiefly responsible for inventing laser cooling and trapping and using these techniques to make the first Bose-Einstein condensate.

While Einstein would surely have been delighted to have witnessed the fulfilment of his 1925 prediction,

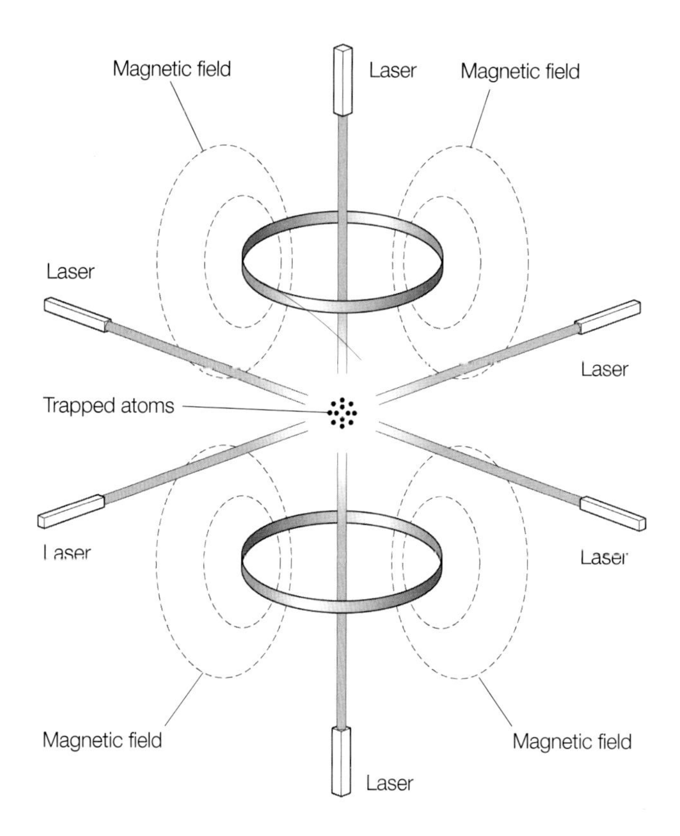

Fig. 7: Laser cooling and trapping was a technique developed in the 1990s, which allowed the creation of the first Bose-Einstein condensate in 1995, 70 years after its prediction by Einstein. Six lasers slow down, and therefore cool, rubidium atoms, and the magnetic fields keep them trapped.

he would most likely have been deeply disturbed by the experimental tests of the 'thought' experiment he published with Podolsky and Rosen in 1935. Entitled "Can the quantum-mechanical description be considered complete?" the EPR paper was yet another attempt by Einstein to show that quantum mechanics was an unsatisfactory theory.

The basic idea was summarized by him in a question he posed in 1933 to Léon Rosenfeld, the physicist who previously described Bohr's discomfort at Einstein's 1930 photon-in-a-box experiment. Einstein asked Rosenfeld:

What would you say to the following situation? Suppose two particles [such as electrons or photons] are set in motion towards each other with the same, very large, momentum, and that they interact with each other for a very short time when they pass at known positions. Consider now an observer who gets hold of one of the particles, far away from the region of interaction, and measures its momentum; then, from the conditions of the experiment, he will obviously be able to deduce the momentum of the other particle. If, however, he chooses to measure the position of the first particle, he will be able to tell where the other particle is. This is a perfectly correct and straightforward deduction from the principles of quantum mechanics; but is it not paradoxical? How can the final state of the second particle be influenced by a measurement performed on the first, after all physical interaction has ceased between them?

In terms of Heisenberg's uncertainty principle, Einstein was saying that if one were to use a precise measurement of one particle's *momentum* to determine the precise momentum of the other, then this must increase the uncertainty in the *position* of the second particle. If, conversely, one were to determine the first particle's position precisely, this must increase the uncertainty in the momentum of the second particle. (Recall that the uncertainty in momentum multiplied by the uncertainty in position of a particle must always exceed a known constant.) And these changes must take place instantaneously, through some kind of faster-than-light signalling.

Which is exactly what Bohr, in his response to the EPR paper, argued to be true. Bohr believed in nonlocal reality: the two particles really do 'cooperate' in a sort of conspiracy forced on them by the nature of physical reality. Schrödinger promptly dubbed the new notion 'entanglement.'

It was completely unacceptable to Einstein. He believed in local reality: "on one supposition we should, in my opinion, absolutely hold fast: the real factual situation of the system S_2 is independent of what is done with the system S_1, which is spatially separated from the former," he wrote in his seventieth birthday response to Bohr in 1949. In a private letter to Born he said he simply could not credit the existence of "spooky actions at a distance."

For a number of years the issue remained in the realms of speculation. Einstein died in 1955 and with him died the chief opponent of Bohr's view of quantum theory, which had become known as the 'Copenhagen interpretation' after the location of Bohr's research centre. Then in 1966 the physicist John Bell devised a theorem which suggested experiments with two photons that could be used to test for the existence, or not, of entanglement. With advances in laser and photon detection technology, it was now possible to produce two photons from one source and then monitor the photons' degree of correlation as they flew apart at high speed. Experiments began in the 1970s and continued for some years in various laboratories, notably in the work of Alain Aspect and his group. By the 1990s it was clear that entanglement was a real phenomenon. Quantum theory and the Copenhagen interpretation had triumphed once again; entanglement became a growth area in physics. Though Einstein's 1935 'thought' experiment had proved very influential in triggering real experiments, it has not given the answer he must have expected and hoped for.

The picture of physical reality that quantum theory gives us is, however, far from easy to grasp and full of mysteries—beginning with the very concept of wave-particle duality that unavoidably implicates the observer in the question of what is measured: a wave or a particle. In quantum theory physical reality is dependent on the observer and hence on human consciousness. Einstein's lifelong adherence to a physical reality independent of human beings is the one that most of us instinctively feel to be true. For science, this view of reality has been immensely productive from the ancient Greeks up to the present day. Maybe that is why Max Planck felt such extreme reluctance to introduce the quantum into physics in 1900 and then resisted its spreading influence in 1905 and after. For Planck had accepted—unlike Einstein (or so it would appear from Einstein's final years spent in search of an ultimate theory)—the fact that "Science cannot solve the ultimate mystery of Nature," as Planck once remarked. "And it is because in the last analysis we ourselves are part of the mystery we are trying to solve."

Einstein's Scientific Legacy

Philip Anderson

In the half century since his death, the mystique surrounding Einstein has created a cult that in my view starts clever physics students by the thousand off in entirely the wrong direction. The cult makes Einstein into the embodiment of a 'pure' theorist, a genius so brilliant that he snatches his ideas from thin air and achieves revolutionary advances solely by the exercise of mathematical reasoning. Students aspire to imitate this great figure, especially in these days of stasis and frustration in high-energy physics; they imagine that if they can achieve deep erudition in higher mathematics, however difficult and esoteric, it is bound to lead them to new insights into physics.

To be sure, there have been physicists who have approximated the ideal of the Einstein cult. Among the most famous was Prince Louis de Broglie, who wrote down the first equation for matter waves in the 1920s—thereby earning a Nobel prize—without, as he bragged, ever leaving his *cabinet*. But afterwards de Broglie did nothing further—he took no part in discovering how to apply his waves to real physics. Another Nobel winner, Dirac, also came close to the ideal—his greatest achievement seems to have been almost entirely motivated by mathematical convenience.

But Einstein himself, during his most productive years, was by no means one of those types. In his "Autobiographical Notes," published when he was 70, he recalled how in his student days in Zurich he could easily have acquired "a sound mathematical education." Instead, "I worked most of the time in the physical laboratory, fascinated by the direct connection with experience." He also wrote: "during my student years I was not much interested in higher mathematics. Mistakenly it seemed to me that one could easily squander one's entire energy in one of its remote parts. I thought in my innocence that it sufficed for a physicist to understand elementary mathematical concepts clearly...and that the remainder consisted of subtleties that were fruitless." Considering that these statements represent his attitude during his *annus mirabilis*, 1905, he cannot have been too "mistaken" or "innocent" then!

Einstein maintained throughout his life a healthy interest in experimental physics and during most of his career a realization that physics is, at bottom, an experimental science. Only in his last years did he become mired in the project to unify the quantum theories of gravity and of electromagnetism, a project about which he had only mathematical speculations, no empirical input at all. Of course he failed.

Einstein in his garden in Princeton, June 1941.

It happens that much of Einstein's early work, especially that around 1905, was about what is now called condensed matter and atomic physics, so that I may be at least as competent to understand its implications as physicists of other stripes. In fact, arguably, Einstein was the first quantum theorist of condensed matter, my specialty.

But let me first turn to relativity, Einstein's most famous work. Relativity, too, did not come out of thin air but was stimulated by experimental fact.

Einstein's 1905 paper on relativity was entitled "On the electrodynamics of moving bodies," and its argument is, in modern terms, very simple: if one is to dispense with the ether and assume that the electromagnetic field propagates freely in space without reference to its source, then space must transform under translations in the same way that Maxwell's equations do, which was already known to Lorentz. That is, the equations must not change in a moving system. Einstein was not the first to suggest dispensing with the ether, nor did he invent the transformation; but he was the first to put the two together and follow through the consequences. He did not need the Michelson-Morley experiment to tell him that Maxwell's equations were invariant, there was tons of evidence for that; he cites electromagnetic induction—that it does not matter if you move the wire or the magnet. (The reams that have been written on the Michelson-Morley experiment and its 'proof' or 'disproof' of relativity are, generally speaking, waste paper. Einstein, being a courteous soul, referred kindly to the experiment two years later.) Soon Einstein went on to treat the electromagnetic field as a gas of free photons, which makes little sense if it is tied either to its source or to an ether.

Einstein in 1935.

To emphasize my point, with relativity in essence Einstein started from Maxwell's equations, that is to say from a century of painstaking experiment (culminating in Hertz's demonstration of radio waves) that Maxwell and Lorentz had summarized for him in these equations. He did not just stick his head in the air and think; one can even argue that with his antipathy to 'groupthink,' his making the best of his intuition, and his deep study of physical reality, he was just connecting up the experimental dots.

What of general relativity? Here the experimental basis is even clearer: it is the equality of gravitational and inertial mass, a fact which centuries of astronomical observation, not to mention ballistic practice and elementary physics experiments, had made perfectly evident. What Einstein realized was that this 'obvious' fact was not a tautology—but something which required explanation. It was a delightful surprise to me to learn that his first paper on the 'equivalence principle,' which became general relativity in the end, was already written in 1907, and that in it he had already derived the gravitational red shift; so that his physical understanding preceded his mathematical understanding of 1915 achieved after eight years of hard labour on the calculating treadmill. Perhaps it was his pride in those eight years of mathematics which led him to insert "mistakenly" in the above-quoted "Autobiographical Notes"—but how, even if he had been truly mistaken, would he have known that it was differential geometry he must study in order to crack general relativity and not, say, Diophantine equations?

Now I turn to the rest of Einstein's work in 1905. In the *annus mirabilis* Einstein wrote five papers which I will discuss not in order of their publication but in order of what I regard as their increasing importance. The first is Einstein's doctoral thesis on the molecular theory of liquids, at which he had been labouring for several years, and it is hardly revolutionary. Closely related was his theory of Brownian movement, my second choice from the five papers, a well-crafted piece of classical statistical mechanics noteworthy for the fact that Einstein did not hear of Brown's 1820s measurements until later in the year, so only his reprise published in 1906 is called "On the theory of the Brownian movement." This second paper, "On the movement of small particles suspended in stationary liquids required by the molecular-kinetic theory of heat," is neat, but, contrary to the mythology, most contemporary physicists already believed in atomic theory and in the statistical mechanics developed by Boltzmann, Maxwell and Gibbs; and there were already pretty good estimates of atomic radii and Avogadro's number. So Einstein's paper convinced only laggards and refined the numbers.

The third and fourth papers, according to my estimate, were on relativity and the brief derivation from it of the equation $E = mc^2$. It is always considered strange and unfair that Einstein did not receive the Nobel prize for relativity but for his fifth paper on the quantum theory. I am in the minority who believe that the Swedes were spot on. Apparently, Einstein himself was another member of this minority, since this was the only paper of his entire career which he called "revolutionary." So much for another myth about him, that he did not like the quantum theory!

The fifth paper—the most important of the five in my view—"On a heuristic point of view about the creation and conversion of light," is based on yet another experimental fact, Planck's formula for the black-body radiation. This paper contains the photon hypothesis: that light behaves like a gas of free particles called photons, each of which carries exactly energy $h\nu$, ν being the frequency of the light wave. Here is the true birth of the modern quantum theory, in that Einstein invents here the idea of a quantum field, the representation of the world which we now recognize as universal and fundamental—that every bit of matter or energy is made up of particle-like objects described by quantized fields. He accepts also, in essence, that light obeys Maxwell's wave equation—so here, 20 years before Schrödinger's wave equation, we have the idea of wave-particle duality: a truly revolutionary concept.

Of course, even Einstein was not infinitely wise. It took him nearly 20 years to arrive at the final form of the theory of the Bose-Einstein field, with the help of a little paper from Bose that arrived unsolicited 'over the transom' from Dacca in 1924. During most of that time he was equivocal about the real existence of the photon (though the name was not given until 1926 and not by Einstein), backing away from his wonderfully bold 1905 hypothesis in an uncharacteristically timid way.

Nonetheless, the photon is really there in his 1905 paper. And what is more to the point for me as a 'solid-stater,' Einstein immediately drew from this idea not just one but three practical experimental confirmations. The famous one concerns the photoelectric effect. The photon gives up its energy to a single electron, which leaves the metal after overcoming the surface barrier, with a fixed residue of energy dependent only on the frequency, not on the intensity, of the light. There was already qualitative evidence for this, and it was soon to be completely confirmed. The photoelectric effect is still one of the most valuable experimental probes of condensed matter, and in every experiment today we exploit Einstein's concept of "one photon, one electron." Hence Einstein's paper is often referred to as being "on" the photoelectric effect, which it was not.

His second experimental confirmation concerns photo-ionization of gases. It had been observed that gases could only be ionized by light of a sufficiently high frequency (sufficiently blue), which was different for each gas; and this too is an obvious consequence of the photon hypothesis. Finally, Einstein notes that fluorescent light—light stimulated by light—is basically always of lower frequency than the stimulating light, a fact which was already known and was called Stokes's rule; which again can be explained only by the quantum hypothesis.

In 1907, still several years before the next paper by anyone else on the quantum theory, Einstein came up with another quantum explanation of a puzzling solid-state effect: the drop in the specific heats of solids at low temperatures. Specific heat is the amount of energy necessary to raise the temperature by one degree Kelvin. Classical statistical mechanics models the atoms in a solid crystal as connected by little springs, in other words as a collection of 'harmonic oscillators,' each of which contributes exactly the same amount to the specific heat. But using the quantum hypothesis, when the energy of one quantum of oscillation is bigger than the thermal energy, the oscillator can no longer respond as readily. Einstein, to get a rough-and-ready answer, assumed all the little oscillators had the same frequency, which could be estimated from the elastic stiffness of the crystal and the mass of the atoms. He thus achieved a remarkably accurate account of the general behaviour of simple solids using the same energy distribution for these 'phonons' that he had already assumed for photons. (Of course, the word phonon had yet to be coined.) We still approximate phonons in certain cases by 'Einstein phonons.'

In 1911, Einstein joined Nernst on the committee of the legendary Solvay Congress at which he, Nernst, Planck, Peter Debye and a few others persuaded the world of

Einstein and his second wife Elsa at the University of Commerce, Tokyo, Japan, 1922.

theoretical physics to take the quantum theory seriously. The stage was set for the next era of the theory, which belonged really to Bohr. But Einstein's contributions to my subject certainly did not dry up. I shall mention three of them.

One which I know well, because I used it in my thesis in 1949, is the relationship between the absorption and emission of light, the so-called 'A' and 'B' coefficients, which Einstein wrote down in 1917 in a paper that much later gave rise to the laser. In due course this relationship was given the fancy name 'Fluctuation-Dissipation theorem.' I may possibly have been the first to put it to practical use. I wanted to derive the absorption spectrum of a gas for microwaves; but it was much easier just to follow the molecules' motions and calculate the light they emitted—and then use Einstein's relation. Generalizations of these methods are now a workhorse of 'condensed matter' theory.

Another example, which is a bit notorious, is the Einstein-de Haas effect. Einstein predicted that rotating a metal would cause it to have a magnetic moment of a certain size. When his friend Wander Johannes de Haas measured the magnetic moment, the size came out just right—but it turned out that de Haas had 'selected' his data, undoubtedly because of Einstein's already great reputation. The later discovery (in 1925) that the electron is spinning—and Dirac's relativistic explanation of why, in 1932—made it clear that the real effect should have been twice as big as that measured by de Haas; and indeed it is. (The gradual rise of this coefficient from 1 to 2, over the years, is an amusing study—though no reflection on Einstein.)

But probably the most famous of all Einstein's later observations, stimulated by Bose's paper of 1924, was that a gas of free bosonic atoms would condense into what turns out to be a superfluid. It is unfortunate—or perhaps fortunate for Fritz London, who first noted the relevance of Einstein's observation—that by the time the superfluidity of liquid helium was discovered in the late 1930s, Einstein was embroiled in his fruitless speculative search for a unified field theory, and so he played no role in this work.

To conclude, Einstein was important to the two greatest revolutions in physical science since Galileo and Newton. The lesser one, relativity theory, he created almost alone. But in the one which in my view is by far the greater, the quantum theory, he played a unique and indispensable role. Out of his work came, as a by-product, the profession of which I am proud to be a member. Today, 'my' Einstein bids fair to be hijacked by the string theorists—and I resent that.

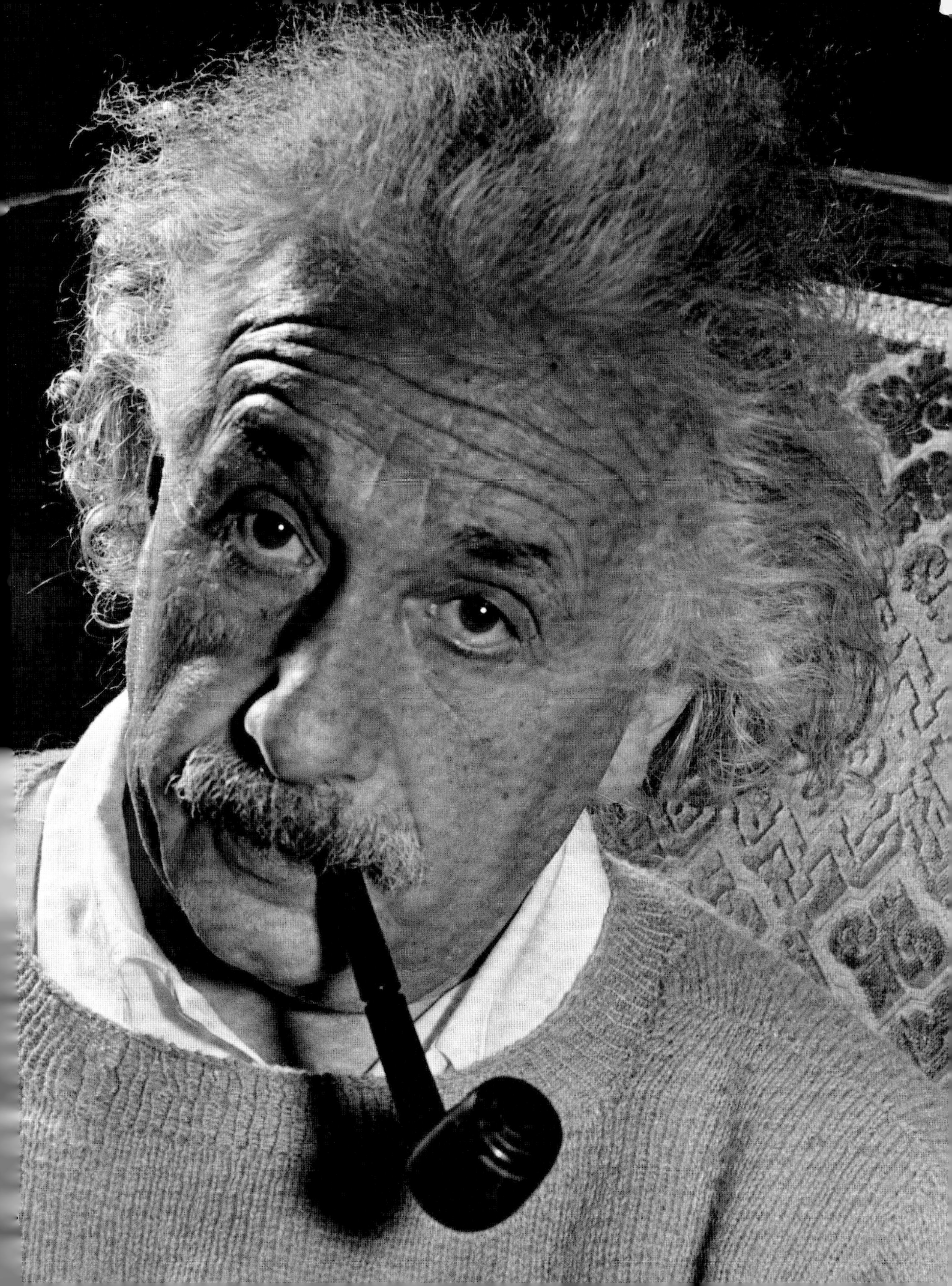

8. The Most Famous Man in the World

"I never understood why the theory of relativity with its concepts and problems so far removed from practical life should for so long have met with a lively, or indeed passionate, resonance among broad circles of the public... I have never yet heard a truly convincing answer to this question."

Einstein, preface to a biography of himself by Philipp Frank, 1942

"Einstein on verge of great discovery; resents intrusion" ran a headline in the *New York Times* on 4 November 1928. Ten days later, the paper declared: "Einstein reticent on new work; will not 'count unlaid eggs'." Soon, to escape all the attention and concentrate on his work, Einstein left Berlin and went into hiding; he stayed alone in a friend's house throughout the winter, cooking for himself, "like the hermits of old" as he wrote to his old friend Besso. On 10 January 1929, when another old friend Max Planck at last delivered Einstein's new paper, "On the unified field theory," to the Prussian Academy in Berlin on his behalf, there was feverish interest from the world's press. Einstein was said to have solved the "riddle of the universe." Telegrams came from all over the world requesting information and 100 journalists were held at bay by the academy until the publication of the six-page paper on 30 January. A delay of three weeks between delivery and publication was routine for a scientific paper, but given the unheard-of public speculation on its content the academy decided to up the usual print run to 1000 copies. All instantly sold out and three further printings of 1000 copies each were hastily arranged—a record for the academy's proceedings.

On 3 and 4 February, the *New York Times* and *The Times* in London carried a full-page article by Einstein, "The new field theory," which discussed mainly relativity and then attempted to explain his latest idea of "distant parallelism" (which he would soon quietly abandon). The *New York Herald Tribune* went one better and printed a translation of the entire scientific paper, including all the mathematics. For technical reasons to do with telex, the mathematical symbols had to be transmitted from Berlin in a code agreed with some cooperative physicists at Columbia University in New York, who decoded the formulas and reconstructed the equations for the newspaper—apparently without inaccuracy, though one wonders just how many readers of any newspaper, no matter how serious, were in a position to judge.

Einstein delivers his Nobel lecture, Gothenburg, 1923. (He was awarded the prize in 1922 for the year 1921.) Fourth from left in the front row is King Gustav V of Sweden.

Einstein and Winston Churchill at Chartwell, Churchill's country home, 1933. They strongly agreed on the menace of Nazi Germany.

But the most extraordinary response, surely, came when the six pages of Einstein's paper were pasted up side by side in the windows of Selfridges department store in London for the benefit of shoppers and passers-by. "Large crowds gather around to read it!" an amazed and amused Arthur Eddington wrote on 11 February in a letter to Einstein.

When Einstein and his wife appeared as the personal guests of Charlie Chaplin at the premiere of his film *City Lights* in Los Angeles in 1931, they were cheered as they battled their way slowly through the frantically pressing crowds—on whom the police had earlier threatened to use tear gas—and the entire theatre rose in their honour. A somewhat bemused Einstein asked Chaplin what it all meant. "They cheer me because they all understand me, and they cheer you because no one understands you," quipped Chaplin.

"The speed with which his fame spread across the world, down through the intellectual layers to the man in the street, the mixture of semi-religious awe and near hysteria which his figure aroused, created a startling phenomenon which has never been fully explained," one of Einstein's many biographers, Ronald Clark, would write half a century later.

Of course it had all started with Eddington and the eclipse, some ten years earlier. On 6 November 1919, his expedition's observations had been presented as the sole item on the agenda of a joint meeting of the Royal Society and the Royal Astronomical Society in London. So important was the occasion that the audience was virtually a roll-call of the greatest names in British physics, astronomy and mathematics. Alfred North Whitehead, mathematician and philosopher, who had come specially from Cambridge, described the scene:

> *The whole atmosphere of tense interest was exactly that of the Greek drama. We were the chorus, commenting on the decree of destiny in the unfolding development of a supreme incident. There was a dramatic quality in the very staging—the traditional ceremonial, and in the background the picture of Newton to remind us that the greatest of scientific generalizations was now, after more than two centuries, to receive its first modification. Nor was the personal interest wanting; a great adventure in thought had at length come safe to shore.*

First the astronomer royal, Sir Frank Dyson, who had launched the whole eclipse enterprise in 1917 despite Britain's being in the midst of a great war with Germany, outlined the course of the two expeditions to West Africa and Brazil and the essentials of the photographic plates of the eclipse on 29 May 1919, and declared that "A very definite result has been obtained that light is deflected in accordance with Einstein's law of gravitation." Then Eddington speaking for the African expedition and a fellow astronomer for the South American expedition presented the observations in detail. Finally the presidents of the two scientific societies supported Dyson and Eddington. Sir J. J. Thomson, discoverer of the electron and president of the Royal Society—and thus the holder of the chair once occupied by Newton—said of relativity:

> *this result is not an isolated one; it is part of a whole continent of scientific ideas... This is the most important result obtained in connection with the theory of gravitation since Newton's day, and it is fitting that it should be announced at a meeting of the Society so closely connected with him... If it is sustained that Einstein's reasoning holds good...then it*

Einstein with his second wife Elsa (seated, fifth from left) in a formal group portrait in Singapore, 1922, where they were guests of the local Jewish community.

is the result of one of the highest achievements of human thought.

But Thomson did allow that relativity was hard to grasp, even for mathematically adept physicists. While the meeting was dispersing, an astronomer came up to Eddington and said, "Professor Eddington, you must be one of the three persons in the world who understands general relativity." When Eddington demurred, his colleague persisted "Don't be modest, Eddington," and received the reply: "On the contrary, I am trying to think who the third person is." Once this story got around, it became something of a game among press commentators to guess the correct number of people in the world who really understood relativity.

The following day, 7 November, *The Times*, along with its main headline concerning King George V's call to observe the first anniversary of the armistice with Germany on 11 November, carried a report of the scientific meeting and a substantial leader article about it. The headlines—"Revolution in science," "New theory of the universe," "Newtonian ideas overthrown"—were followed by the subheadings "'Momentous pronouncement'" (a further quote from Thomson) and "Space 'warped'." The leader concluded that, "It is confidently believed by the greatest experts that enough has been done to overthrow the certainty of ages and to require a new philosophy of the universe, a philosophy that will sweep away nearly all that has hitherto been accepted as the axiomatic basis of physical thought." Across the Atlantic, having been scooped by the British paper, on 9 November the *New York Times* gave the announcement even more prominence with six headlines. These led with "Lights all askew in the heavens," followed with "Stars not where they seemed or were calculated to be, but nobody need worry," and finished up—in a first US reference to the incomprehensibility of the new theory—with the smaller headlines "A book for 12 wise men" and "No more in all the world could comprehend it, said Einstein when his daring publishers accepted it." This *New York Times* report, and *The Times* report, marked the appearance of Einstein's name in the two newspapers.

It was also more or less the inauguration of relativity among American physicists. Without an Eddington to

Einstein and Elsa aboard the SS *Kitano Maru*, during their travels to the Far East in 1922.

Einstein with Queen Elisabeth of Belgium.

legitimize the theory, and especially given its purely theoretical basis, relativity was destined to fall on sceptical ground in the United States. Albert Michelson, the first US winner of the Nobel prize in physics—mainly for the Michelson-Morley experiment to detect the ether—was unconvinced by relativity when he died in 1931. "The supposed astronomical proofs of the theory, as cited and claimed by Einstein, do not exist," stated Charles Lane Poor, professor of celestial mechanics at Columbia University. He was reminded of Lewis Carroll: "I have read various articles on the fourth dimension, the relativity theory of Einstein and other psychological speculation on the constitution of the universe; and after reading them I feel as Senator Brandegee felt after a celebrated dinner in Washington. 'I feel,' he said, 'as if I had been wandering with Alice in Wonderland and had tea with the Mad Hatter.'" An engineer named George Francis Gillette was so outraged that he called relativity "the moronic brain child of mental colic...cross-eyed physics...utterly mad...the nadir of pure drivel...and voodoo nonsense." By 1940, said Gillette, "relativity will be considered a joke... Einstein is already dead and buried alongside Anderson, Grimm, and the Mad Hatter." However, such fierce criticism from professionals only fed the interest of the US public in the new scientific sensation. When Einstein lectured at the American Museum of Natural History on his first visit to the US in 1921, the New York City police had to be called to quell their first "science riot," such was the crush to gain entry.

At the Swedish Academy in Stockholm, the sceptics were out in force, as we might perhaps have expected. Einstein was first nominated for the Nobel prize in physics in 1910 and nominated again by Nobel laureates in every year thereafter except 1911 and 1915. In 1920, after the public announcement of the confirmation of general relativity, his name dominated the nominations. Still, a leading member of the physics committee could assert: "Einstein must never receive a Nobel prize even if the whole world demands it." By 1922, the pressure was impossible to resist and the deferred prize for 1921 was given to Einstein. Yet, as we know, it was awarded not for relativity and not for light quanta, but for "his discovery of the law of the photoelectric effect"—with the stipulation that his Nobel lecture must be on this subject, as is customary with such lectures. (In the end, such was the public interest, including that of the Swedish king, by the time Einstein gave the lecture in July 1923, he spoke on relativity.)

To return to the inventor, as opposed to his controversial theory, in November 1919 Einstein himself was a complete blank to the English-speaking press. *The Times* report referred only to "the famous physicist, Einstein" and did not give even his first name or the fact that he worked in Berlin. But the very incomprehensibility of relativity made newspapers all the hungrier for information about the personality behind it. And once he was contacted and interviewed Einstein rapidly proved to be a deft, witty and eminently quotable popularizer of his own ideas. Even he, though, could not transform relativity into more than a tantalizing concept for the non-mathematically minded. In his little book entitled *Relativity* intended for the general reader he still felt the need for some, admittedly high-school, mathematics. Planck once remarked: "Einstein believes his books will become more readily intelligible if every now and again he drops in the words 'Dear reader'"—a remark which an amused Einstein liked to quote. In his later years, besieged with requests from journalists and the public, he was reduced to telling his secretary to give casual enquirers the following explanation of relativity: "An hour sitting with a pretty girl on a park bench passes like a minute, but a minute sitting on a hot stove seems like an hour."

His public image was undoubtedly helped by the fact he enjoyed communicating his ideas. Never, in reading Einstein's writings—whether scientific or otherwise—does one feel that he is taking refuge in obscure language; there is always a striving for the utmost possible clarity. And this attitude appears to have been reflected in his conversation too. Never, or at least hardly ever, did he lay down the law as a recognized authority and expect others to defer; rather he showed a genuine humility and willingness to learn from those he respected. While he was not by nature a great teacher,

Einstein answering hundreds of questions posed by reporters when the SS *Belgenland* docked at New York, 1930.

because he had too many original ideas, he had a gift for making himself understood. For press reporters an Einstein interview always made excellent copy.

Among the first to discover this—and the very first person to write a book about him—was the Berlin literary journalist Alexander Moszkowski. Based on meetings in the period 1919-20, his *Conversations with Einstein* catches its subject on the cusp of fame. Einstein's physicist friends, such as Max Born, were appalled that he had agreed to the book and tried to stop it from appearing; but even they eventually admitted it had merit. Though Moszkowski was no scientist, he made up for his scientific gaucheness with his vivid picture of Einstein's personality, which recalls the high-spirited debates of the Olympia Academy in Bern two decades earlier. "We all know," said Moszkowski, speaking obviously of Einstein's German audiences, "that no iron curtain marks the close of Einstein's lectures; anyone who is tormented by some difficulty or doubt, or who desires illumination on some point, or has missed some of the argument, is at liberty to question him. Moreover, Einstein stands firm through the storm of all questions." On the day of one of their conversations, Einstein had come straight from a lecture on four-dimensional space which had provoked a raging debate. "He spoke of it not as of an ordeal that he had survived, but as of a refreshing shower."

So among the more important reasons for Einstein's extraordinary rise to fame in the early 1920s has to be his personal accessibility to the public. Another reason must be the historical moment. In 1919, just after the end of a bloody conflict without precedent, English scientists had confirmed the theory of a German scientist. For peoples desperate for peace, this cooperation between very recent former enemies seemed almost providential. Eddington told Einstein on 1 December: "For scientific relations between England and Germany this is the best thing that could have happened." Moszkowski chimed with this view and felt an even wider resonance: "The mere thought that a living Copernicus was moving in our midst elevated our feelings. Whoever paid him homage had a sensation of soaring above Space and Time, and this homage was a happy augury in an epoch so bare of brightness as the present."

Yet another reason for Einstein's fame arises from the capacious name of his theory: relativity. It had been given not by Einstein but by Planck and others as early as 1906. Einstein was never quite comfortable with it—he would have preferred a name to do with invariants or invariance—but around 1911 he acquiesced and began to use it in his own writings. "The term 'theory of relativity' is an unfortunate choice," wrote the well-known physicist Arnold Sommerfeld much later. "Its essence is not the relativity of space and time but rather the independence of the laws of nature from the viewpoint of the observer. The bad name has misled the public into believing that the theory involves a relativity of ethical conceptions, somehow like Nietzsche's

Einstein is feted with a motorcade by the city of New York on his first visit to the USA, 1921.

Einstein and his wife Elsa at a tea-drinking ceremony in Kyoto, Japan, 1922.

Beyond Good and Evil." For better or worse, over time, in many non-scientific minds relativity would come to mean pretty much what one wanted it to mean and more or less the opposite of what Einstein meant by it—something like 'Everything is relative.'

There is also the interest in relativity and religion to consider. Science versus religion always stirs the public. In answer to a question from the archbishop of Canterbury, when he visited England in 1921, Einstein said firmly that relativity was a purely scientific matter and had nothing to do with religion. But he was not always so definite and wrote much on the relationship of science and religion. Many others, including some physicists, did see a connection. Erwin Schrödinger, for instance, speaking of special relativity in a lecture given the year after Einstein's death, said:

> *It meant the dethronement of time as a rigid tyrant imposed on us from outside, a liberation from the unbreakable rule of 'before and after.' For indeed time is our most severe master by ostensibly restricting the existence of each of us to narrow limits—70 or 80 years, as the Pentateuch has it. To be allowed to play about with such a master's programme believed unassailable until then, to play about with it albeit in a small way, seems to be a great relief, it seems to encourage the thought that the whole 'timetable' is probably not quite as serious as it appears at first sight. And this thought is a religious thought, nay I should call it* the *religious thought.*

Last but by no means least in trying to account for Einstein's fame must be his defiant individualism. His views were his own, regardless of whether they made him popular or not with authority or with the mass, in Germany, in his adopted home in the United States or elsewhere.

We shall see this again and again in the next chapters. People all over the world sensed the integrity in Einstein's public statements and in the man when they met him in person—whether they were Chaplin,

Winston Churchill or just an ordinary school teacher or plumber—on his travels in Europe, North and South America, Palestine and the Far East during the 1920s. George Bernard Shaw saluted him at a grand public dinner in London as "a kind of great man who did not create empires but universes, whose hands are not sullied by the blood of a single human being." This was long before Einstein became known after the Second World War as the sage with crazy hair who wore sweat shirts instead of suits and shoes without socks. His integrity made people curious about him and his ideas, if not always comprehending. In Tokyo, in 1922, at the traditional Chrysanthemum Feast in the imperial gardens to celebrate the union between the Japanese imperial family and the people, it was the guest of honour Einstein to whom the Japanese participants chose to pay attention, not the empress or the prince regent or the princes—whether the royal family liked it or not. The German embassy in Tokyo reported to Berlin that, "the approximately 3000 participants... because of Einstein totally forgot what the day signified. All eyes were turned on Einstein, everyone wanted at least to shake hands with the most famous man of the present day."

The person at the centre of all this unique attention was not averse to being the world's most famous living man, though he did not court the role. And he was decidedly willing to exploit his fame shrewdly to further the ethical causes he believed in. But he did not claim to have fathomed his appeal. "Why is it that nobody understands me, yet everybody likes me?" Einstein asked a *New York Times* interviewer in 1944. Five years later, on his seventieth birthday, he gave a typically acute scientific analogy for his celebrity in a letter to his friend Born, who had known him since before he became famous: "I really cannot understand why I have been made into a kind of idol. I suppose it is just as incomprehensible as why an avalanche should be triggered off by *one particular* particle of dust, and why it should take a certain course."

Opposite: Einstein with Charlie Chaplin at the Los Angeles premiere of Chaplin's film *City Lights*, 1931.

Einstein speaks at a high-profile fund-raising dinner at the Savoy Hotel, London with Lord Rothschild (centre) and George Bernard Shaw (right), 1930.

9. Personal and Family Life

"What I most admired in him as a human being is the fact that he managed to live for many years not only in peace but also in lasting harmony with a woman—an undertaking in which I twice failed rather disgracefully."

Einstein, letter to the Besso family on the death of Michele Besso, 1955

Not long after Einstein died in April 1955, his stepdaughter Margot, who had been close to him for decades and was there at the end, wrote a letter to a very old friend of his, Hedwig Born, Max Born's wife. "He left this world without sentimentality or regrets," Margot told her. Her physicist husband later commented on this simply: "With his death, we, my wife and I, lost our dearest friend."—the final words of *The Born-Einstein Letters.*

There is an inescapable fact about Einstein contained in these brief remarks. It is this: he meant a huge amount to his wide circle of friends and to the world at large—but to himself, once he could no longer work, the world, even his friends, meant comparatively little. "I am truly a 'lone traveller' and have never

Einstein's first wife, Mileva Marić, and their two sons Eduard and Hans Albert, 1914, the year of the couple's separation.

belonged to my country, my home, my friends, or even my immediate family, with my whole heart," he stated in "The world as I see it," a sort of credo published when he was about 50.

We might feel this to be a bleak view of human possibility, yet there is an honesty to it, which seems to have been shared by geniuses in many fields in their later years. Einstein, in his long and turbulent life as well as at his death, was practically never sentimental and seems to have suffered exceptionally little from regret. "Nothing tragic really gets to him, he is in the happy position of being able to shuffle it off. That is also why he can work so well," his second wife Elsa (mother of Margot) confided to a woman friend soon after the trauma of losing her first daughter Ilse to disease. "Albert's devotion to science takes precedence over his emotional commitments. It is a defining truth," stresses the Einstein papers editor Robert Schulmann in his article in this book on Einstein's love letters to his first wife Mileva Marić.

In the early years, around 1900, there can be no shadow of doubt that Einstein loved Mileva, whatever the growing strength of his commitment to physics. In one letter to her he says: "When I'm not with you I feel as if I'm not whole. When I sit, I want to walk; when I walk, I'm looking forward to going home; when I'm amusing myself, I want to study; when I study, I can't sit still and concentrate; and when I go to sleep, I'm not satisfied with how I spent the day." Some months later he writes: "You are and will remain a shrine for me to which no one has access; I also know that of all people, you love me the most, and understand me the best." Later still, after she becomes pregnant, he tells her: "When you're my dear little wife we'll diligently work on science together so we don't become old philistines, right? My sister seemed so crass to me. You'd better not get that way—it would be terrible. You must always be my witch and street urchin. I want to see you so badly. If I could only have you for a little while!"

But there are hints, even in these affectionate remarks, of future trouble. Albert would always be active; Mileva was essentially passive. He needed love; but he would find it hard to return love. Above all, he wanted to do science; and Mileva valued "human happiness" (she told her best friend) more than "any other success." Add to this the implacable opposition of Einstein's mother to her son's marriage to Mileva, and one can understand why it quite quickly went sour.

Einstein with his son Hans Albert and his grandson Bernhard, 1932.

Their first child, the daughter whom they called Lieserl, was born in January 1902, a year before they married. Einstein was still jobless but had the prospect of the position at the Swiss Patent Office that he at last secured in June that year. But for him to become a civil servant in Swiss bourgeois society with an illegitimate child, he and Mileva knew to be impossible. Most probably Lieserl was adopted in Mileva's native Serbia some time in 1903, after the Einsteins had married and set up home in Bern, though no one can be certain in the absence of more than meagre circumstantial evidence.

Hans Albert, their next child, was born the following year, and their second son, Eduard, in 1910. Throughout this period, until Einstein was promoted to a full professorship in Prague in 1911, the couple were short of money. Einstein supplemented his salary with private tuition, and it is from some of his students that we get a picture of the domestic conditions *chez* Einstein. David Reichinstein recalled entering Einstein's room and finding his teacher "calmly philosophic...with one hand rocking the bassinet in which the child was lying (his wife was at work in the kitchen). In his mouth Einstein had a bad, a very bad cigar, and in his other hand was an open book. The stove was smoking horribly. How in the world could he bear it!" (On one occasion this stove almost asphyxiated him as he lay down on a couch to sleep; he was saved only by a chance visit from an academic friend, Heinrich Zangger, who quickly opened the windows.) Einstein's first PhD pupil, Hans Tanner, gives an even more graphic account:

> *He was sitting in his study in front of a heap of papers covered with mathematical formulae. Writing with his right hand and holding his younger son in his left, he kept replying to questions from his elder son Albert who was playing with his bricks. With the words: 'Wait a minute, I've nearly finished,' he gave me the children to look after for a few moments and went on working. It gave me a glimpse into his immense powers of concentration.*

By now, of course, Einstein had published the great papers of 1905 and was about to embark in earnest on the work that led to general relativity. He and his wife had apparently ceased to discuss physics. A few claims have been made—for which there is some slight evidence—that in the early years of the marriage Mileva had played an important intellectual role, and even that Einstein should have shared the authorship of the relativity paper with his wife (like the collaboration between Pierre and Marie Curie). But when one looks at Mileva's educational record and correspondence, such a role seems extremely unlikely. For sure, she was an unusually bright student in mathematics and physics at her school in Zagreb at a time when this subject was very difficult for a girl to pursue, however her first choice of study at the Swiss Polytechnic in Zurich had been medicine not physics; she failed the final exam there in 1900 mainly on her weak mathematics; and her single surviving reference to the 1905 papers in her letters to her best friend is merely glancing: "the papers he has written are already mounting quite high." Some four years on Mileva told the same friend: "He is now regarded as the best of the German-language physicists, and they give him a lot of honours. I am very happy for his success, because he really does deserve it; I only hope and wish that fame does not have a harmful effect on his humanity." No Einstein scholar has felt able to take seriously the claims for an intellectual partnership with Mileva, which are advanced chiefly by Serbian writers, presumably for patriotic reasons.

Over time Einstein seems to have become convinced that mathematics and physics were subjects not

Einstein sails with his step-daughter Ilse and her husband Rudolf Kayser (who wrote a biography of Einstein).

suitable for women, though it should be said that in 1919 he strongly supported the outstanding mathematician Emmy Noether in her struggle to break the all-male grip on German universities. When in the 1920s a young woman physicist in Berlin, Esther Salaman, who was attracted to Einstein, confessed to him that she probably lacked the creativity required for theoretical physics, he replied: "Very few women are creative. I should not have sent a daughter of mine to study physics. I'm glad my wife doesn't know any science; my first wife did." But was Marie Curie not creative, asked Salaman? "We spent some holidays with the Curies," said Einstein. "Madame Curie never heard the birds sing!" Speaking to his second wife, he was blunter about Curie, saying that she was "very intelligent but has the soul of a herring, which means that she is poor when it comes to the art of joy and pain." After Curie's death he publicly extolled her "strength of character and devotion" but even then could not refrain from adding that she had "a curious severity unrelieved by any artistic strain." Yet despite his comment on the lack of joy in birdsong, Einstein also told Salaman: "I'm not much with people, and I'm not a family man. I want my peace. I want to know how God created the world." She was not quite convinced, though; she sensed in his conversation a strong desire for human contact but also a fear of it. "His voice was a protection of his inner self rather than an expression of it. He felt kindly towards people, but not intimate with them, not even sure of them."

Einstein relaxes at Palm Springs, California, 1932.

It is scarcely difficult to find evidence of male chauvinism in Einstein's life. While this may have been typical of the university milieu of his time, Einstein took it further than some of his friends, such as Besso. The brief section on "Women" in *The Quotable Einstein*, a collection of his best remarks, is a dispiriting part of an otherwise stimulating book. But what is also abundantly obvious from Einstein's letters and his several love affairs is that he enjoyed the company of women—"Einstein was a flirt," conclude the authors of *The Private Lives of Albert Einstein*, on sound evidence—and that some of these female friendships were close. None of the women he knew, even Mileva, ever wrote publicly against him, despite the temptations offered by his fame. What is more, Einstein's perhaps most loyal supporter, who became one of the executors of his will, was a woman: his secretary Helen Dukas, who served him for three decades and beyond, fiercely protecting his private life from scrutiny after his death. Einstein was in no sense a misogynist, though he had relatively little regard for female mental powers. Overall, there is an element of contradiction in his attitudes to women, both in general and as individuals. But what is beyond dispute is that he was not cut out for marriage—as Einstein himself more or less openly admitted.

It was the move to Berlin in 1914 that finally precipitated the break. Mileva soon returned to Zurich with their two sons. There was much bitterness on both sides, though Einstein contrived to maintain contact

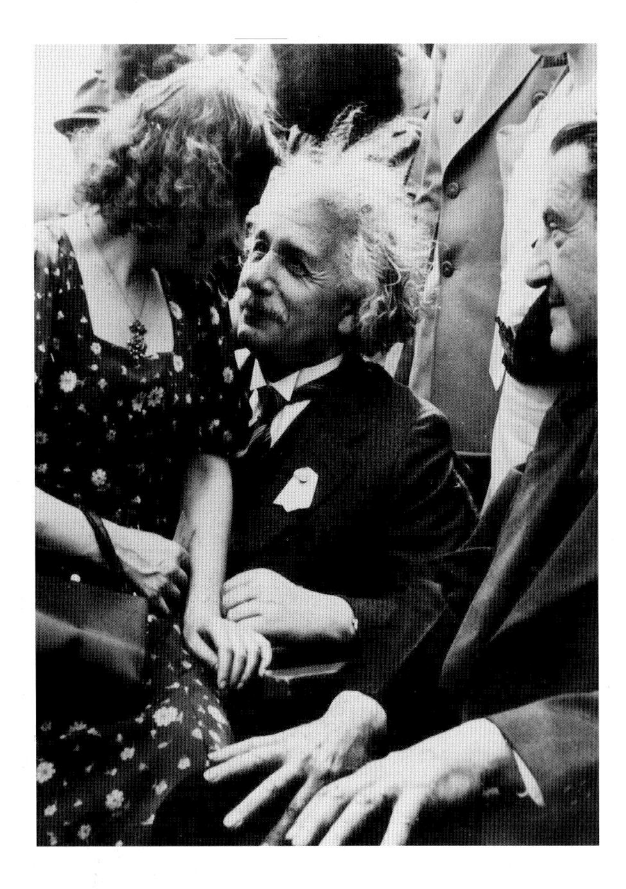

Einstein with his step-daughter Margot and Rabbi Stephen Wise, at the opening of the New York World's Fair, 1939. Einstein wrote a statement that was included in a time capsule which was to be sealed for 4,000 years.

with his sons as far as possible. By 1918 his wife had accepted the separation and divorce proceedings began, mediated by the couple's mutual friends in Zurich, Besso and Zangger. Both parties agreed to a unique proposal: that the money from Einstein's yet-to-be-awarded Nobel prize should be used to support Mileva and the two children. Clearly she retained her original faith in her husband's genius. Einstein, for his part, was obliged to admit adultery. In early 1919—the same year that would shortly see his rise to world fame—the divorce papers were signed. A few months after this, Einstein married again.

Elsa Löwenthal (*née* Einstein), his second wife, was his cousin. Her mother was the sister of Einstein's mother; her father was a first cousin of Einstein's father. She too was divorced and the mother of two girls, Ilse and Margot, living with her in a comfortable apartment in Berlin. Albert had known her as a child in southern Germany and in his student days Elsa would have personified for him all the bourgeois, philistine, *gemütlich* complacency that he rebelled against when he married the brooding Mileva Marić. "She might have passed in her prime as a Valkyrie in an amateur production of Wagner" (from *The Private Lives* again). But when Einstein met her on a visit to Berlin in 1912, he felt differently and they started an affair. Her primary interests were domestic and hardly intellectual; though she had a taste for literature (she gave public readings of poetry), she was totally ignorant of science. In marrying her, Albert tacitly accepted that his mother Pauline had been right all along. Over the years after 1919, until her death in Princeton in 1936, Elsa would be his devoted companion and protector on his world travels, relishing his fame unashamedly—and lending a theatrical air of absurdity to the endless photographs of the great man—while tolerating his unconcealed affairs with other women.

Following the divorce, Einstein's relationship with his first wife eventually improved. He periodically visited her and the two children in Zurich, and they exchanged several hundred letters in the last two decades of her life up to her lonely and lingering death in Zurich in 1948. Though it was Mileva who would raise Hans Albert and Eduard, their father was keen to contribute to their education. But for neither son would the results of his interventions be happy.

Hans Albert decided to follow a scientific career like his father; in due course he became a hydraulic engineer with an international reputation. However an applied science held no appeal for Einstein. There was an angry confrontation when Hans Albert was 15. "I think it's a disgusting idea," said his father. "I'm still going to be an engineer," insisted Hans Albert. Whereupon Einstein strode off saying he never wanted to see his elder son again. After some time, when he had cooled down, Mileva was able to bring them together again, and eventually Einstein came to take pride in his son's achievements. But they clashed again badly over Hans Albert's marriage, Einstein reiterating similar arguments against it to the ones that his parents had

Einstein and his wife Elsa, on his first trip to the USA, 1921

Einstein with his son-in-law, Rudolf Kayser.

used against his own marriage. When Hans Albert and his family emigrated to the United States in 1938, following his father and step-mother, they settled finally at Berkeley in California, far from Princeton. There were no more open conflicts, but father and son kept their distance, despite sharing the same two recreations, music and sailing. In his professional career Hans Albert found his household surname an embarrassment and spoke as little as possible about his famous father.

Eduard Einstein's life was sad and even tragic. As a teenager, unlike his elder brother he showed glints of his father's genius, in literature and music rather than science. At first Einstein encouraged him, but then he decided his son was becoming pretentious. Soon Eduard was drawn towards psychiatry and Sigmund Freud, who his father knew slightly and whose ideas he distrusted. While studying psychiatry at university in Zurich Eduard became seriously depressed, suffered a breakdown and announced that he hated his father. In 1932, he was admitted to a mental institution in Zurich for the first of many stays intended to treat him for schizophrenia. Einstein's friends Besso and Zangger urged him to look after Eduard. But Einstein refused on the grounds that the condition was inherited from Mileva's family and one could do little to combat the "secretory causes." As he had written sceptically for Tagore two years earlier: "should the lawfuless of events, such as unveils itself more or less clearly in inorganic nature, cease to function in front of the activities in our brain?" Apparently for Einstein this philosophical belief in determinism exonerated him from taking personal responsibility for his sick son.

He visited Eduard only once in the institution, in 1933, a few months before he left Europe forever to live in America. Thereafter he would not even write to him, while continuing to support his treatment financially and to worry about him. In 1952, while thanking his son's faithful carer Carl Seelig, he said of Eduard: "He represents virtually the only human problem that remains unsolved. The rest have been solved, not by me but by the hand of death." A year before his own death he excused himself to Seelig for not writing directly to his son: "There is a block behind it which I am unable to analyse fully. But one factor is that I believe I would be arousing painful feelings of various kinds in him if I made an appearance in whatever form." Eduard outlived his father by ten years, dying in the mental institution in 1965.

In writing about such personal tragedies, it is only too easy to rush to judgement without knowing the full facts. When he wrote those first words about his second son, Einstein himself had been fatally ill for four years. In this period he had shown himself to be far from indifferent to his family while looking after his bedridden sister Maja in Princeton during a protracted illness. After the war she had intended to return to Switzerland but had been incapacitated by a stroke. Every evening her brother read to her from good literature, classic and modern. "Strangely enough, her intelligence had hardly suffered from the advanced illness, even though towards the end she could hardly talk audibly anymore," he told a friend after her death in 1951. "Now I miss her more than can be imagined." He was also exceedingly busy in other ways. Year after year he assisted complete strangers with requests for help while supporting numerous good causes and conducting a worldwide campaign against the erosion of international peace and national civil rights by the forces of political intolerance (as we shall see in later chapters). *And* he continued with his scientific work on the unified field theory.

Yet still there is a paradox. How could Einstein have treated his own children so aloofly, while making himself accessible to thousands of people he did not know at all? In the final analysis, it seems, Einstein cared more about humanity than about individual human beings. He distrusted those feelings which he had called, in his "Aubiographical Notes," the "merely personal."

Overleaf: Paper cutouts of his family created by Einstein and pasted into a copy of the children's book *Peterchens Mondfahrt*, sent as a present to the young son of some friends, 1919. The silhouettes, which took him two hours to make, depict Einstein himself (labelled A), Elsa (E), Ilse (I) and Margot (M). It is signed 'A. E. pinxit', Latin for 'painted by A. E.'

Einstein with his step-daughter Margot and his wife Elsa and a friend.

Peterchens

Felix [illegible] den Kleinen
Von den sämtlichen Einsteinen
Dass er ja nicht [illegible] vergisst
Jedes hier verewigt ist.

I. M

A. E. pinxit.

Mondfahrt.

Einstein's Love Letters

Robert Schulmann

We all recognize Albert Einstein, the elderly sage of Princeton. But what do we know about the inner landscape of Einstein's life? His openness in publicizing scientific and political thoughts was not matched by similar revelations in the personal sphere. Little could he anticipate that the rediscovery after his death of his love letters to Mileva Marić would afford readers a privileged glimpse into those inner worlds of his youth, which he fiercely guarded during his lifetime. Yet on closer examination, the yield is even richer. Supplemented by other correspondence from that time, the love letters sound the emotional leitmotifs that are to course through his entire life.

They tantalize not only for their references to an intellectual and romantic relationship, but also for what they leave unsaid. While any correspondence can plumb only some of the depths of a relationship, there are a number of clear asymmetries that mark these letters. Only ten of Mileva's survive, as against 44 from Albert's hand. Equally disproportionate is the contrast between his increasingly authoritative voice and her more halting presence. The language of their correspondence—German—is Albert's native tongue, and his ideas and emotions tumble onto the page in an often dramatic cascade of expressive phrases. On the other hand, the reserved Mileva, a native Serbian speaker, often seems unable to get her sparely crafted words in edgewise. There is, in addition, an ebb and flow to the letters. Mileva, the 22-year old, dominates the correspondence at its inception in 1897. She reassures and comforts her 19-year-old classmate. Most importantly, she reinforces Albert's passion for science and music, though claims that Mileva contributed significantly to Albert's groundbreaking papers of 1905 are not supported by these letters.

A relationship that begins with talk of books consulted, experiments undertaken, and hypotheses examined, blossoms into a romance, fuelled in part by shared musical evenings. A strict protocol of address is observed. Until 1900, the formal "Sie" is used. After that Mileva and Albert employ the informal "Du" and invent affectionate names for one another. She is "Doxerl" (Dollie), he is "Johonzel" (Johnny), diminutives which they exchange over endless cups of coffee. In a letter of May 1901, Albert's mention of an exciting new experiment for generating cathode rays by ultraviolet light is followed by his reaction to the startling news that Mileva is pregnant. The sequence is significant. Albert's devotion to science takes precedence over his emotional commitments. It is a defining truth that will persist for the remainder of his life.

After a five-year courtship, the couple marry in 1903. By this time Albert has found his intellectual and emotional footing and has assumed an almost magisterial role in the relationship. Mileva is now the junior partner. The camaraderie and bravado which had marked their earlier relationship recedes. No longer are they sustained by a youthful pride in their own resilience and a rejection of bourgeois accommodation. The two former bohemians are now confronted with the practical demands of marriage, including a wrenching decision to give up a daughter born out of wedlock. As Albert begins to make his way professionally, the relationship deteriorates even further. In the winter of 1909–10 Mileva writes to her best friend that success does not leave her husband, a newly minted professor in Zurich, much time for his wife. To the query whether she is "jealous of science," she responds: "...what can you do? One gets the pearl, the other the jewel box." Mileva is trapped in a nostalgic depression, while the supple Albert partakes of the insights and emotions of a variety of worlds while keeping his distance from all of them.

Other contemporary sources reveal a similar ambivalence in Albert's dealings with women. One of the most revealing is a letter he addresses to Pauline Winteler, in whose home he boarded while attending secondary school in Switzerland. Writing in the spring of 1897, at a time when he has already developed an interest in Mileva, Albert dramatically announces that he must break off his schoolboy friendship with Pauline's daughter Marie, as it is "strenuous intellectual work and the study of God's Nature [that] are the reconciling, bracing, yet relentlessly austere angels that will lead me through all of life's troubles." While a certain amount of practical calculation no doubt plays a part, it is striking how here, as gradually with Mileva at university, Albert is able to sublimate his emotional life in a wilful distancing of himself from engagement with the individuals around him.

Not just in his physics was Einstein an adherent of determinism. Drawing primarily on the philosopher Arthur Schopenhauer, whose aphoristic writings he avidly read and continued to cite throughout his life, he also applied it in the personal sphere. Repeatedly, at moments of personal moral crisis, especially in his relations with women, Einstein sought refuge in a philosophical denial of free will, moral agency, and thus personal moral responsibility. It was this that lay at the heart of his ruthless pursuit of a divorce from Mileva in 1919, of the flippant reference to his irresponsibility as a father for his frequently callous treatment of his two sons, and of his affairs with a number of women while married to his second wife Elsa, from whom he hid nothing. After she died in

1936 in Princeton, Einstein pursued at least two more liaisons until his own death in 1955.

Taking stock of his emotional life on reaching the age of 50, Einstein attempted, unsuccessfully, to reconcile his impersonal commitments with personal aloofness. A "passionate interest in social justice and social responsibility has always stood in curious contrast to a marked lack of desire for direct association with men and women.... Such isolation is sometimes bitter but I do not regret being cut off from the understanding and sympathy of other men... I am compensated for it in being rendered independent of the customs, opinions and prejudices of others and am not tempted to rest my peace of mind upon such shifting foundations."

Einstein at the piano, 1933.

Einstein and Music

Philip Glass

"If I were not a physicist, I would probably be a musician," Einstein is reported to have said in 1929. "I often think in music. I live my daydreams in music. I see my life in terms of music... I get most joy in life out of my violin."

Einstein on the beach—on Long Island, New York.

Einstein took up the violin at the age of six, and played it for most of his life until, in the last few years, he gave it up and played only the piano. He was modest about his abilities as a musician. Then again, he also said he wasn't that good a mathematician, and I treat that statement with some scepticism. A music critic, seemingly confused about the identity of the performer he was reviewing, once commented that Einstein's violin playing was "excellent, but he does not deserve his world fame; there are many others just as good."

Whatever his musical talent, it is clear that music was very important to him. Above all, he was Einstein the Scientist—certainly the greatest since Newton. But for me, he was also Einstein the Dreamer, a type of personality one often sees in the arts but which one associates less frequently with scientists, and here perhaps is one of the keys to this uniquely creative scientist.

Einstein brought to bear something very different when approaching scientific problems. It was through his famous 'thought' experiments on relativity, such as imagining himself sitting on a beam of light travelling through space, that he managed to envision the answers to the problems he was tackling. But then he had real work to do, because the thought experiments preceded the mathematical proof, and he had to set about developing new tools of description for this purpose. This story of how a personal vision became the cornerstone of modern physics is certainly one of the greatest intellectual adventures of modern times. Special relativity was a hypothesis based on an act of imagination: the scientist entering the realm of the dreamer, the poet, the artist. That's what interests me most about Einstein.

When I was growing up in Baltimore in the 1940s, Einstein was the first celebrity from the world of science. I read his books, and was very affected by the birth of the nuclear age which this gentle man had indirectly instigated. At the University of Chicago in the 1950s I studied mathematics and philosophy along with the normal liberal arts curriculum of the college. Einstein remained a personal inspiration even after I chose to concentrate on music, studying in New York and Paris. Strangely enough, as I was establishing myself in music back in New York, one of the jobs I took to support myself was as a plumber, another career Einstein once said he envied.

Einstein with the musicologist Alfred Einstein. The two Einsteins were not related, but they had met in Berlin after having been sent each other's mail by mistake, and they remained in contact after both moved to the USA.

In 1974 the theatre director Robert Wilson and I were trying to come up with an idea for a major new musical work on which we would collaborate. We wanted it to be about a famous person, someone who would resonate with the population at large. We used to meet regularly for lunch to discuss our ideas, but at first we were unsure which figure to pick. Bob proposed Charlie Chaplin, who I thought would be too difficult, and Hitler, who I declined. I countered with Gandhi, who Bob rejected (Gandhi became the subject of my second opera). Then Bob suggested Einstein. With the image of the great dreamer in mind, I agreed enthusiastically, and *Einstein on the Beach* was born.

I suppose Einstein also seemed an appropriate choice of subject because the type of work we were creating was rather radical. We were living in a post-Newtonian world, which needed Einstein's new tools to explain it, and this was a post-modern opera; in fact I was never sure whether to class it as an opera at all, since it seemed to break many of the normal rules of the genre. It had an abstract structure and no actual plot; the spoken texts had no explicit connections to Einstein, and their phrases were repeated to fill the allotted time, the whole work taking more than five hours. Our thinking was that everyone already knew who Einstein was; the audience would complete the work by determining the meaning for themselves. I eventually settled on the term 'portrait operas' to describe this and my next two works, *Satyagraha* and *Akhnaten*. These two, along with *Einstein*, together formed a trilogy—each of them a musical/theatrical portrait of a historical figure who changed the world through the power of ideas and not through violence.

Although the work is not a traditional biography by any means, Einstein is still its central theme, and the staging included plenty of visual references to him. In the original production, the dancer and choreographer Lucinda Childs appeared with an 'Einstein' pipe, and all the cast members were dressed in the Einstein 'uniform'—sneakers, white shirt, braces and baggy pants. One character, resembling Einstein right down to the hairstyle, played the violin all through the opera, sitting halfway between the orchestra and the stage, thereby acting as an intermediary between the music and the action. At the start of act one there was a scene with a train, referring to another of Einstein's relativity thought experiments, while other visual themes—spaceships, elevators, gyroscopes, watches, clocks—all referred back to our subject.

Einstein on the Beach toured France, Italy, Germany, Yugoslavia, Holland and Belgium in 1976, and was then performed twice at the Metropolitan Opera

House in New York, where we had something of a triumph. It proved to be a turning point in my career and my life. *Einstein* itself has been revived several times over the years, and later I composed other science-related projects: a score for a film about Stephen Hawking, *A Brief History of Time*, and a recent opera about Galileo Galilei.

Back in 1976, despite the sold-out houses at the Met, *Einstein on the Beach* was very expensive to stage and in the short term we were saddled with what we called 'the Einstein debt.' I returned to driving a taxi, another means of supporting myself in those days. I vividly recall the moment shortly afterwards when a passenger noted my name on the taxi/limo licence (required to be displayed by the New York City Taxi and Limousine Commission), and said: "Young man, do you realize you have the same name as a very famous composer?" And this too, thanks to Einstein.

10. Germany, War and Pacifism

"A funny lot, these Germans. To them I am a stinking flower, and yet they keep putting me into their buttonhole."

Einstein, in his travel diary in South America, 1925

As a boy Einstein never wanted to play with toy soldiers or to become a soldier. When he was just short of 17 he relinquished his German nationality mainly to avoid compulsory military service. On becoming a Swiss citizen in 1901, he was not called up because the authorities found him medically unfit. As an adult he never at any time wore military uniform and the only weapon he ever handled seems to have been a sword required by Habsburg imperial protocol when he was a professor in Prague, which he could not avoid donning for a ceremonial swearing in. Throughout the First World War, when he lived in Berlin, he made no secret of his opposition to the use of military force, while remaining on friendly terms with scientific colleagues who had joined the fight in earnest. He hated war and for most of his life—though not, significantly, during the Nazi regime in Germany and of course not during the Second World War—he was a militant pacifist.

"That a man can take pleasure in marching in fours to the strains of a band is enough to make me despise him," Einstein wrote in 1930 in "The world as I see it." He continued trenchantly:

> *Heroism on command, senseless violence, and all the loathsome nonsense that goes by the name of patriotism—how passionately I hate them! How vile*

Einstein (face turned away) visiting the battlefields of the First World War.

Einstein at the Royal Albert Hall, London, in October 1933, where he was one of many prominent speakers (others included Ernest Rutherford and the British politician Austen Chamberlain) at a gathering in aid of the Jewish Refugee Fund. At the time Einstein was living in exile in England, after fleeing Nazi Germany and just before moving to America.

A heliogravure print by Rose Weiser of Einstein during his Berlin years.

and despicable seems war to me! I would rather be hacked in pieces than take part in such an abominable business. My opinion of the human race is high enough that I believe this bogey would have disappeared long ago, had the sound sense of the peoples not been systematically corrupted by commercial and political interests acting through the schools and the press.

At least in this respect, he thought, the masses were more trustworthy than the intellectuals. Ordinary people were not as prone as academics and professional thinkers to the "psychoses of hate and destructiveness," he wrote in an open letter to Sigmund Freud not long before the Nazi seizure of power, published as a pamphlet called *Why War?* in 1933. In making this judgement on his own milieu, he most probably had in mind his disturbing experiences of the patriotic behaviour of his scientific colleagues in 1914–18. Certainly he must have been thinking of the German academic attacks on himself and relativity as the work of 'Jewish science' in the 1920s and early 30s.

The psychosis declared itself in October 1914, two months after the outbreak of war, when 93 leading Germans from the world of the arts, humanities and sciences enthusiastically signed what would, during and after the war, become a notorious document. (In 1919, at the time of the eclipse observations, *The Times* noted approvingly that Einstein had *not* been one of the signatories.) Published in the leading German dailies, it was also translated into ten languages throughout the world. Entitled a "Manifesto to the cultured world," it protested "the lies and defamations with which our enemies are trying to besmirch Germany's pure cause in the hard life-and-death struggle forced upon it." It denied that Germany had started the war, defended Germany's breach of Belgium's neutrality, dismissed the stories of atrocities committed by Germany's troops as fabrications and proclaimed that Germany's cultural legacy—Goethe, Beethoven and Kant were mentioned by name—and its current militarism were as one. Among the scientists who signed, in addition to the two conservative nationalists and Nobel laureates Wilhelm Wien and Philipp Lenard (of whom more in a moment), Einstein must have been pained to see the names of three of his friends, Max Planck, Walther Nernst and Fritz Haber.

As a Swiss citizen he had not been asked for his signature. But he now decided to make his first-ever public political statement by signing a counter-manifesto drafted by a well-known German physician and physiologist Georg Friedrich Nicolai, whose patients included the imperial family. The "Manifesto to the Europeans," though it openly rejected the 'Manifesto of the 93,' did not analyse the causes of the war and attribute guilt. Instead in deliberately restrained language it urged educated people everywhere to try to "create an organic unity out of Europe... Should Europe, too, as Greece did earlier, succumb through fratricidal war to exhaustion and destruction? For the struggle that is raging today will hardly leave a victor but only vanquished behind." Nicolai circulated the

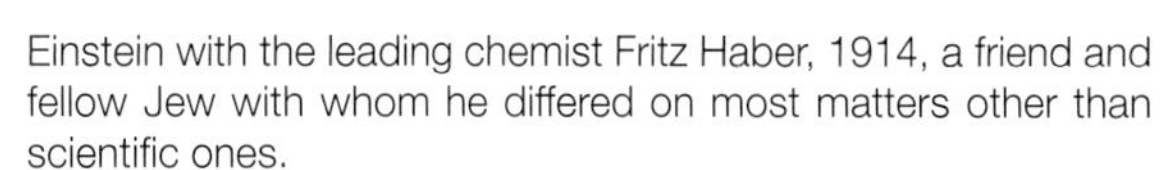

Einstein with the leading chemist Fritz Haber, 1914, a friend and fellow Jew with whom he differed on most matters other than scientific ones.

Einstein with Max Liebermann (left), the German painter who like Einstein was persecuted by the Nazis, and the sculptors Renée Sintenis and Aristide Maillol, Berlin, 1930.

counter-manifesto among the staff at Berlin University in late 1914. Many indicated agreement (some, including Planck, were having regrets about signing the nationalist manifesto)—but only two others apart from Einstein, one of whom was an unknown, were willing to sign the appeal. A distressed and isolated Nicolai gave up and the counter-manifesto did not appear in print until 1917—and then in Zurich not Berlin. Nicolai's career was ruined by it; ultra-nationalists regarded him as a traitor and succeeded in having him banned from teaching in 1920.

Meanwhile, among the patriots, Wien sent round his own circular. It called upon his academic colleagues to avoid in future quoting scholars from the enemy camp in Britain, even in footnotes, unless such quotations were indispensable. Einstein's colleague and correspondent Arnold Sommerfeld (the future teacher of Heisenberg) declared himself pleased to sign the appeal in a letter to Wien written on—of all inappropriate days—Christmas Day 1914, while German and Allied troops were fraternizing in a truce in the trenches. Two decades later, under Hitler, not even references to the work of Jewish physicists, especially Einstein's—unless they were derogatory—would be permitted. But in the Nazi case this was by government fiat with serious penalties for transgression. What is particularly discouraging about German scientific self-censorship in the First World War is that it was entirely voluntary.

The national self-delusion was perfectly encapsulated in a vignette from Einstein the following year. After every meeting of the Berlin University Senate, he related, laughing aloud, all the professors would meet in a restaurant and "invariably" the conversation would begin with the question: "Why are we hated in the world?" Then there would be a discussion in which everyone would supply his own answer while "most carefully steering clear of the truth."

This account is from the diary of the writer and pacifist Romain Rolland, who won the Nobel prize for literature in 1915. Einstein had come to see Rolland in Switzerland that September in search of a fellow spirit, while he was on a visit to his wife and children in Zurich. When Rolland asked him if he dared to voice his anti-war views to his German friends in Berlin and argue with them, Einstein said he did not. "He confines himself to asking a lot of questions—as Socrates did—in order to upset their peace of mind," Rolland noted. "He adds: 'People don't like that very much.'"

By now, writes Einstein's German biographer Albrecht Fölsing, "Most physicists were...part of the war effort: the younger ones, as junior officers, were in the meteorological service, the artillery, or the chemical warfare industry; the older ones were in their laboratories, perfecting killing machines." Among the juniors was Max Born, who was sent as a scientific assistant to the artillery inspectorate in Berlin—which had the unexpected advantage of bringing him close to Einstein for the first time.

Romain Rolland, writer and pacifist, at his Swiss home, 1936.

The most chilling example of the scientific dedication to war came from one of Einstein's older friends, the chemist Haber, in whose institute he kept an office. The Haber-Bosch process for synthesizing ammonia from nitrogen and hydrogen is a staple of high-school chemistry courses, for which Haber received the Nobel prize in 1918 (though it was not presented until 1919, after the war). The first commercial plant for ammonia production came on stream at the end of 1913 for making artificial fertilizers. However within a year the process was being used to make explosives too; indeed without Haber's process, the German army would have run out of shells as early as 1915. This was clearly a vital contribution to the war effort, but Haber had much wider ambitions to help his country. The army asked him to develop poison gas. In April 1915 Haber supervised the world's first chemical warfare attack, using chlorine, on French troops at Ypres in Belgium; 5000 died and 10,000 suffered severe damage to their lungs. A week later, after a stormy argument with him, Haber's wife shot herself with his army pistol—at least partly out of horror at her husband's lethal role. But Haber's war fever was unstoppable. Even after the war, in the early 1920s, he and his institute worked in total secrecy on developing pesticides as potential chemical weapons. One of the substances was known as Zyklon B. During the Second World War, it was used in the death camps to murder millions, including friends and distant relatives of Haber, who was Jewish. Fortunately for him, he had died of a heart attack in 1934 soon after being expelled from Nazi Germany. Fifty years later, his son, a historian of science, wrote: "In Haber... [the High Command] found a brilliant mind and an extremely energetic organizer, determined, and possi-

Einstein arrives in New York aboard the SS *Belgenland*, 1930. Two years later, he left Germany for good, settled in the United States in 1933, and never left it.

bly also unscrupulous." One might be tempted to compare Haber with Albert Speer, Hitler's architect and armaments minister in the Second World War, another cultured technician corrupted by war.

Einstein could hardly have been unaware of Haber's military activities; it was Haber who had physically supported a weeping Einstein on the way back from the railway station in Berlin when he separated from his wife and family in the summer of 1914. The two men were close during the First World War, despite disagreeing on whether it was just. However the preoccupations of theoretical physics, and in particular general relativity, were undoubtedly an escape for Einstein's mind throughout the war years. His absorption must have helped him overlook the more bellicose aspects of friends like Haber and Nernst. Nor was he absolutely uninvolved in war work himself. Despite his pacifism, he did a little consulting, not with great success, on the design of aircraft wings, and became involved as a patent examiner with Anschütz, a company making gyrocompasses for use in U-boats, with whom in the 1920s he himself would design a better gyrocompass. For whatever reasons, in public Einstein kept quiet about the perversion of science into military technology during the First World War.

But when the officers of the Berlin Goethe League approached him in the autumn of 1915 for a contribution to a patriotic book, he did not restrain himself. "The best minds from all epochs are agreed that war is one of the worst enemies of human development, that everything should be done to prevent it," he wrote. He called for a political organization in Europe to stop European wars "in the same way as the German Reich now rules out a war between Bavaria and Württemberg." These comments were permitted in his contribution, "My opinion of the war," to *The Country of Goethe 1914–1916: A Patriotic Album.* Not so some of his other comments, which were regarded as too mocking and subversive, such as his suggestion that instead of having patriotic shrines it would be better to worship pianos or bookcases. Also censored was this forceful statement: "How close a person or a human organization is to me depends solely on how I judge their intentions and abilities. The state, to which I

Le Petit Journal

ADMINISTRATION — 5 CENT. — SUPPLÉMENT ILLUSTRÉ — 5 CENT. — ABONNEMENTS

26me Année — Numéro 1.278

DIMANCHE 20 JUIN 1915

Victimes de leur propre barbarie

Soldats allemands asphyxiés par les gaz qu'ils avaient lancés contre les Russes et qu'un coup de vent rejette sur leurs tranchées

"Victims of their own barbarity:" headline on the cover of a French magazine, 1915. It depicts German soldiers overcome when a gust of wind blew back towards them the poison gas they had tried to use on their Russian enemies. Einstein's friend Fritz Haber was closely involved in developing gas warfare.

belong as a citizen, plays not the slightest role in my emotional life; I regard a person's relations with the state as a business matter, rather like one's relations with a life assurance company." In November, at the very moment that he was labouring fanatically on general relativity, Einstein took time out to argue with the Goethe League about these censored passages, but eventually he had to give in.

The war ground on, gradually worsening for the German High Command, as Einstein privately knew from his contacts with people in power like Haber and Walther Rathenau, then chairman of the electrical engineering giant AEG. Einstein's view of official failings hardened, and he now looked forward to the collapse of the German military machine as the only hope for peace. From Berlin he told Besso in 1917: "I am reminded of the period of the witch trials and other religious perversions. The most responsible people, the most unselfish ones in private life, are often the most solid pillars of dogged stubbornness." He was probably thinking of friends like Nernst, who had lost two sons killed in action, and Planck, one of whose sons had been killed while a second was a French prisoner of war. (In early 1945 Erwin Planck was executed as part of the Nazi revenge for the plot to assassinate Hitler.) Planck, even in private to Einstein, still stoutly maintained at the start of 1918 that Germany was constantly pursuing peace, despite having an increased military superiority. Nonetheless, Einstein continued to find in Planck a noble figure, as well as a leading physicist. When the war was finally over, because of Planck and some other physicists such as Born and Max von Laue, Einstein could not bring himself to desert Germany in what he felt was its hour of need. The first tempting post-war offer came from Hendrik Lorentz in Holland in September 1919. Einstein finally refused it, saying that to leave Germany would be "tantamount to a vile breach of promise given to Planck and also otherwise disloyal. I would blame myself afterwards."

He stuck with this attitude throughout the 1920s. He felt that German theoretical physics was still the best in the world, or at least the equal of its British counterpart. Moreover he had political hopes of the Weimar Republic that took over from the militarists in 1919. The growth of the far-right parties and of anti-Semitism was something he felt able to endure and was willing to combat. Though never emotionally committed to being a German, Einstein's first language was German, his wife Elsa was German and many of his family lived in Germany. Overall he still felt that Germany was the best place for him to live and work, apart perhaps from Switzerland. But a return to Zurich would have meant living in the same city as his divorced wife.

The original trigger for the anti-Semitic campaign against Einstein may have been his rise to world fame in 1919–20. It was his public pacifism and his forthright political pronouncements that fuelled the abuse, though—including his commitment to the Zionist cause in 1919 and after (which we shall come to in Chapter 12 on Zionism).

Lenard, one of the signatories of the 1914 manifesto, gave the campaign prestige as a Nobel laureate. His was the work on the photoelectric effect that Einstein had interpreted in terms of light quanta in 1905. Lenard, however, rejected Einstein's interpretation, and attacked it, along with quite different work by several eminent scientists. Antagonistic by nature, Lenard became a follower of Hitler as early as 1924. "He invented the difference between 'German' and 'Jewish' physics," wrote Born; and after 1933 (along with another Nobel laureate, Johannes Stark), Lenard set about cleansing German science of Jews. In his *German Physics*, published in 1936, he wrote: "In contrast to the intractable and solicitous desire for truth in the Aryan scientists, the Jew lacks to a striking degree any comprehension of truth."

Born—who, though Jewish too, had always felt closer to Germany than had Einstein—foresaw the course of political events more accurately than his friend. In 1921 he told Einstein he sensed "a wholly irreversible accumulation of ugly feelings of anger, revenge, and hatred." In 1922, Einstein's friend Rathenau, now foreign minister of the Weimar Republic, was assassinated. Einstein responded by reducing his involvement in domestic politics. The following year, he himself received a death threat, just after German newspapers had published a fictitious planted account of his visiting the Soviet Union: a red rag to ultra-nationalists. Possibly by coincidence, the death threat arrived on the very day, 7 November 1923, that the then-unknown politician Adolf Hitler attempted his Beer Hall Putsch in Munich. Einstein fled to Lorentz and his friends in Leiden for a while. For the rest of the 1920s, urged by German friends such as Planck, he undertook little political activity in Germany, but he remained passionately committed to pacifism and the need to resist military conscription. While visiting the United States in 1930 he announced that, "If even two per cent of those called up declare that they will not serve, and simultaneously demand that all international conflicts be settled in a peaceful manner, governments would be powerless." This "two per cent" speech quickly became a rallying cry for pacifists everywhere.

It was the political success of Nazism that changed Einstein's mind about pacifism. From long and hard experience he accepted, much earlier than most, that Hitler was bent on war. During 1932, seeing the desperate state of the Weimar Republic, he became fully active in German politics again and with two others urged the formation of an anti-fascist coalition between the Communists and the Socialists in the mid-year elections. The idea proved hopelessly impractical. In fact the Communists and the Nazis were the parties who managed some kind of collaboration, however brutal, in the months before Hitler became chancellor of Germany on 30 January 1933. Einstein, preparing to leave Berlin for another visit to the US in December 1932, realized that a turning point had been reached. As he and his wife Elsa locked up their villa outside Berlin for the winter, he told her: "Turn around, you will never see it again."

His break with Germany finally came while he was in California in early 1933. Shrewd as ever about political realities, Einstein wrote a letter resigning from the Prussian Academy on 28 March, depriving the new government of the chance of stripping him of his membership. The Nazi authorities in Berlin were apoplectic. The craven Prussian Academy immediately bowed to what it thought was expected of it and accused Einstein publicly of "atrocity propaganda" against Germany. In a personal response to Planck, Einstein, while denying this allegation, added prophetically: "But now the war of extermination against my Jewish brethren has compelled me to throw the influence I have in the world into the balance in their favour." Planck, ever the social conservative, replied equating such persecution of the Jews with Einstein's pacifism and refusal of military service: "Two ideologies, which cannot coexist, have clashed here. I have no sympathy with the one or the other." It was a sad and very bitter way to end the long friendship between the two great physicists.

Einstein photographed on the streets of Berlin on 1 December 1932 by a passer-by, named Charles Holdt, who recognized his face. Holdt then sent Einstein this photograph. It is probably the last picture taken of Einstein in Germany, for he left the country a few days later.

Opposite: Einstein seems to have known, on his departure from Germany in December 1932 for a third successive winter in California, that he would never return.

11. America

"We are unjust in attempting to ascribe the increasing superiority of American [scientific] research work exclusively to superior wealth; devotion, patience, a spirit of comradeship, and a talent for cooperation play an important part in its successes."

Einstein, "Some impressions of the USA," 1931

In Einstein's last years, along with the 'Red scare' the craze for television swept the United States. On a Sunday evening in mid-February 1950, Einstein agreed to be the star guest for the première of the NBC television show *Today with Mrs Roosevelt*, for which he was recorded at home in Princeton making a statement for a programme hosted by Eleanor Roosevelt. There was intense interest from the public. On 30 January President Truman had announced that a new weapon, more powerful by far than the atomic bomb, would be built as quickly as possible. Now Einstein—the world's most famous scientist, the man who had first advised President Roosevelt of the need to develop the atomic bomb in 1939, and one of the two or three best-known figures in America—was going to address this momentous subject on television.

His powerful statement, which must have fallen like a moral bombshell in most living rooms across the nation, was unequivocally opposed to the development of the hydrogen bomb. "If it is successful, radioactive poisoning of the atmosphere and hence annihilation of any life on earth has been brought within the range of technical possibilities," he said.

Einstein further warned his adopted country: *The idea of achieving security through national armament is, at the present state of military technique, a disastrous illusion. On the part of the USA this illusion has been particularly fostered by the fact that this country succeeded first in producing*

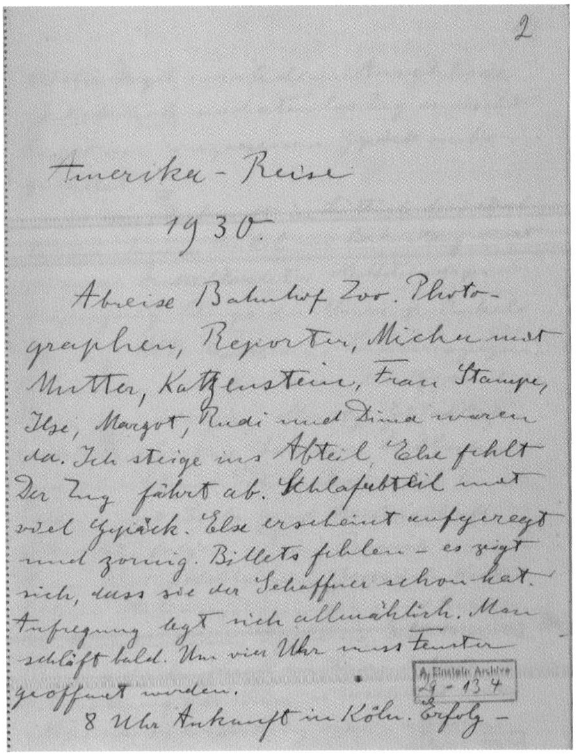
2

Amerika-Reise
1930.

Abreise Bahnhof Zoo. Photographen, Reporter, Michel mit Mutter, Katzensteine, Frau Stampe, Ilse, Margot, Rudi und Dina waren da. Ich steige ins Abteil, Else fehlt. Der Zug fährt ab. Schlafabteil mit viel Gepäck. Else erscheint aufgeregt und zornig. Billets fehlen – es zeigt sich, dass sie der Schaffner schon hat. Aufregung legt sich allmählich. Man schläft bald. Um vier Uhr muss Fenster geöffnet werden.
8 Uhr Ankunft in Köln. Erfolg-

Einstein's travel diary for his second trip to the USA in 1930–31. This first page describes his departure from Berlin's railway station: amid a melee of photographers and reporters, he loses first his wife and then the tickets, before finding both.

Einstein at his home in Princeton, during the recording of his statement for the *Today with Mrs Roosevelt* TV show, 1950.

an atomic bomb... The maxim which we have been following during these last five years has been, in short: security through superior military power, whatever the cost... This mechanistic, technical-military psychological attitude has had its inevitable consequences.

Consequences, said Einstein, not only in foreign policy—which we shall come to in Chapter 13 on Einstein's work for world peace—but also at home in the United States:

Concentration of tremendous financial power in the hands of the military; militarization of the youth; close supervision of the loyalty of the citizens, in particular, of the civil servants, by a police force growing more conspicuous every day. Intimidation of people of independent political thinking. Subtle indoctrination of the public by radio, press, and schools. Growing restriction of the range of public information under the pressure of military secrecy.

US conditions were coming to resemble those in Germany in 1914–18, Einstein might have added (this is what he thought), though he chose to make no reference at all in his statement to his personal experience of war and political intimidation. His advice was: first and foremost, "do away with mutual fear and distrust."

The very next day—probably triggered by the newspaper headlines about Einstein's TV broadcast—the director of the Federal Bureau of Investigation, J. Edgar Hoover, sent a memo to all FBI offices in the country requesting any and all "derogatory information" they had on Einstein. A few weeks later, the government's Immigration and Naturalization Service began its own independent investigation into whether Einstein should have his US citizenship revoked, with a view to eventually deporting him.

For five years, in fact up to Einstein's death in 1955, these two government agencies attempted to find evi-

Einstein, his secretary Helen Dukas (left), and his step-daughter Margot take the oath of American citizenship in 1940.

dence that Einstein was a Communist and, in the case of the FBI investigation, a spy for the Soviet Union. In the atmosphere of McCarthyism, at a time when the state of Texas had approved the death penalty for anyone convicted of membership of the Communist Party, such allegations were commonplace—but even then not about someone as respected and popular as Einstein. So Hoover knew that top secrecy was essential, and thanks to the prevailing atmosphere of fear in government circles there was not a single leak in the press. Some senior FBI officials of the time remained unaware of their agency's campaign against Einstein until the facts emerged in the late 1990s in much-censored government documents. "No one ever accused Hoover of being publicity-shy," writes the investigative journalist Fred Jerome in his eye-opening book *The Einstein File*, published in 2002, "but if it got out prematurely that he was investigating the world's most admired scientist and America's most famous refugee from Nazi Germany, he knew that he and his FBI—and quite possibly the entire United States government—would face a storm of international outrage and derision."

Let us leave the government investigation and revisit it at the end of this chapter. For now it is enough to understand that Einstein's relationship with the United States was always an ambivalent one, from his arrival in Princeton in 1933 until the final departure in 1955, two decades later.

He certainly admired American science, freedom and tolerance. In a 1940 radio broadcast for a network series, *I Am an American*, consisting of interviews with notable immigrants—an item *not* cited against him by the FBI, needless to say—Einstein said:

> *Making allowances for human imperfections, I do feel that, in America, the development of the individual and his creative powers is possible, and that, to me, is the most valuable asset in life... In some countries, men have neither political rights nor the opportunity for free intellectual development. But for most Americans, such a situation would be intolerable. In this country, it has been generations since men were subject to the humiliating necessity of unquestioning obedience.*

But he disliked American hubris, authoritarianism and racism almost as much as their German equivalents. In the 1930s he generally kept these opinions of his host country to himself, but after 1940, when he became a US citizen, and particularly after the beginning of the Cold War in 1946, he chose to speak out strongly in interviews and to support a huge range of American political and semi-political organizations and individuals, famous and unknown, who opposed official policies that curtailed freedom and civil rights.

All his visits to the US prior to settling in Princeton had also mixed politics with theoretical physics—but not American politics. The first trip, in 1921, was controversial because he travelled with Chaim Weizmann, the president of the World Zionist Organization, in order to help raise funds from American Jews for the Zionist project in Palestine, especially for the planned Hebrew

Einstein and Margot during their American citizenship ceremony, 1940. Ever the individualist, Einstein did not wear socks.

University in Jerusalem. An ironic Einstein told his fellow Jew Fritz Haber: "Naturally, I am needed not for my abilities but solely for my name, from whose publicity value a substantial effect is expected among the rich tribal companions in Dollaria." Haber tried strongly to dissuade his friend from going, on the grounds that if Einstein associated himself with Zionism it would backfire against the assimilation of Jews into German society. Although Einstein ignored Haber's advice, he later fell out with Weizmann and his organization—as we shall see in the next chapter.

There were no more Einstein visits to the US until 1930. In the winter of that year, and again in the winters of 1931 and 1932, he accepted an invitation from Robert Millikan of the California Institute of Technology in Pasadena. It was while staying at Caltech in early 1933 that he decided not to return to Germany.

Fifteen years earlier, Millikan had forcefully rejected Einstein's light quanta as an explanation of the photoelectric effect. Now of course, like everyone else, he accepted Einstein's unique scientific status, and he knew that Einstein would be a prize catch for the up-and-coming Caltech. However the two men were poles apart politically, socially and temperamentally. Millikan came originally from an old New England colonial family, though his grandparents had settled in the Midwest as pioneers and his father was a preacher; he himself had grown up on a farm in Iowa. Like many American university administrators of his day, such as A. Lawrence Lowell of Harvard University, Millikan was discreetly anti-Semitic. (He even worried about whether to have Robert Oppenheimer back at Caltech in 1945 after the completion of the atomic bomb project, "because of the large percentage of his fellow racists [*sic*] who are already appearing at the Institute," as Millikan put it.) Einstein's informality and outspokenness—not to speak of his unbuttoned personal appearance—were not at all in tune with Millikan's pompousness and cautiousness. And he was disturbed by Einstein's speeches on pacifism, which he knew would hold even less appeal for the conservative sponsors of Caltech.

A 'socialist' speech that Einstein chose to give to the Caltech students in 1931 must have been particularly unwelcome to Millikan. "Concern for man himself and his fate must always form the chief interest of all technical endeavours," said Einstein, "concern for the great unsolved problems of the organization of labour and the distribution of goods—in order that the creations of our minds shall be a blessing and not a curse to mankind. Never forget this in the midst of your diagrams and equations."

Away from Caltech, returning to the east coast by rail across the continent, Einstein continued to pursue his independent agenda, mixing 'anti-racism,' pacifism and Zionism. Near the Grand Canyon, he visited a group of Indians from the Hopi tribe. He posed for a fine photograph in a magnificent Indian headdress holding a pipe and was given the title, 'the Great Rela-

tive'—yet another spin on the name of Einstein's famous theory. In Chicago, when the train stopped for two hours, he spoke from its rear platform to several hundred supporters of peace and again advocated civil disobedience to military conscription as he had in his famous "two per cent" speech not long before: "It is an illegal fight, but a fight for the real right of the people…when [governments] demand criminal actions from their citizens." Back in New York, Einstein agreed, at the insistence of Weizmann, on the evening of his departure for Europe, to be the guest of honour at a big hotel banquet on behalf of the American Palestine Campaign. More than 1000 participants paid $100 each, in the middle of a great economic depression, to hear him. Even President Hoover, no friend of pacifism or socialism, sent a telegram which received an ovation: "My hope is that your visit to the United States has been as satisfying to you as it has been gratifying to the American people."

Yet the following year, 1932, conservative officials in the State Department of the Hoover administration tried to prevent Einstein from receiving a visa unless he would sign a statement that he was not a Communist or an anarchist. They were acting on a bundle of paranoid unsourced allegations submitted to the government and to the press by a fringe group calling itself the Woman Patriot Corporation. It was led by the well-connected Mrs Randolph Frothingham, who was known for her advice to the Daughters of the American

The Einsteins visiting a group of Hopi Indians near the Grand Canyon, 1931.

A group portrait taken on the White House lawn in 1921, where the Einsteins and the US Academy of Sciences were the guests of President Warren Harding. Einstein and Elsa are in the middle, alongside Harding.

Einstein on a bicycle at the home of friends in Santa Barbara, California, 1933.

Revolution in compiling their blacklist of public speakers. One sentence claimed emphatically that, "Not even Stalin himself is affiliated with so many anarcho-communist international groups to promote...world revolution and ultimate anarchy, as Albert Einstein."

To begin with Einstein thought this was all a joke and wrote a short sardonic riposte for the front page of the *New York Times*: "But are they not perfectly right, these watchful citizenesses? Why should one open one's doors to a person who devours hard-boiled capitalists with as much appetite and gusto as the ogre Minotaur in Crete once devoured luscious Greek maidens—a person who is also so vulgar as to oppose every sort of war, except the inevitable one with his own wife?" But when he and Elsa were called into the US consulate in Berlin and seriously questioned about the allegations, Einstein became furious and threatened to cancel his US visit. The government quickly backed down and a visa was issued. But the unsubstantiated allegations of the Woman Patriot Corporation would find their way, unknown to Einstein, into the FBI's Einstein file and provide useful ammunition for J. Edgar Hoover in the 1950s.

Had Einstein been a normal Nobel laureate in physics, he would have avoided politics altogether when he became an exile at the newly founded Institute for Advanced Study in Princeton in October 1933. That is unquestionably what its influential director Abraham Flexner hoped for from his star physicist. Flexner made extraordinary efforts to restrain Einstein from doing anything but sit and think. For example, when Flexner heard about a New York benefit concert for refugees where Einstein had agreed to play his violin, he tried to stop him by making threatening phone calls to the organizers saying he would fire Einstein from the institute. When the White House invited Einstein to meet President Roosevelt, Flexner intercepted the letter and replied on Einstein's behalf that the professor had come to the US only for scientific work. This was too much for Einstein, who now sent a five-page list of Flexner's misdemeanours to the trustees of the institute along with a threat to resign. As a result, Einstein won perfect freedom for himself—and became the Institute for Advanced Study's greatest asset—but only at the cost of losing all influence in the running of the institute. In January 1934 he and his wife duly dined with the Roosevelts at the White House, which marked the beginning of a significant personal relationship. (Even on this occasion, according to his secretary, Einstein declined to wear socks.)

Given Einstein's later role in the story of the atomic bomb and his attempt to control the growth of nuclear weapons, it is interesting to compare him with the other physicist, also of course Jewish, who became synonymous with nuclear issues in the mind of the post-war American public: Oppenheimer. As director of the Institute for Advanced Study from 1947, Oppenheimer and Einstein were colleagues. But they were "nearly opposites in temperament and outlook" says a recent biographer of Oppenheimer, the physicist David Cassidy. He contrasts them:

> *Einstein the consummate outsider, Oppenheimer the consummate insider; Einstein the loner, Oppenheimer the committee man and administrator; Einstein the constructor of universes without much student assistance, Oppenheimer the founder of a school and facilitator of the work of others; and Einstein the politically engaged defender of civil rights, critic of McCarthy, and supporter of the scientists' movement against nuclear weapons, Oppenheimer the politically removed defender of government policies, victim of McCarthyism, and former head of nuclear weapons development.*

At the height of the hounding of Oppenheimer as a supposed security risk in 1954, Einstein was asked to give a press statement. In private he burst out laughing and said that all Oppenheimer needed to do was go to Washington DC and tell the officials they were fools. But after some reflection, he did compose a statement and read it himself to the press over the telephone. He said: "I can only say I have the greatest respect and warmest feelings for Dr Oppenheimer. I admire him not only as a scientist but also as a man of great human qualities." His first reaction would have been good advice for an outsider like himself, but useless for an insider like Oppenheimer.

His reaction also suggests why the US Army, supported by the FBI, did not give Einstein security clearance to work on the Manhattan Project to build the atomic bomb (in contrast with the US Navy who used his advice on high explosives during the war). It is a strange irony of history that Einstein was the first to alert the US government to the technical possibility of the atomic bomb—in a letter to Roosevelt written in August 1939—but was then banned by the US Army from working on it. A further irony is that Oppenheimer, the scientist put in charge of the project, was particularly close to the Communist Party in the 1930s, whereas Einstein always kept his distance from the party. (He never visited the Soviet Union, despite many invitations.) The precise reason why the army denied Einstein clearance has never been made clear—the 1940 letter in question is the only one in Einstein's FBI file to have vanished—but the most likely explanation is that the army authorities were scared of his unique influence. Both on his fellow physicists and on the American public. Einstein was too independent and fearless a figure to manage, especially given his instinctive pacifism. And from their own point of view the generals may well have been right. Einstein's support for the atomic bomb was entirely predicated on his fear that the Nazi regime would develop such a bomb before the Allies. Had he become aware of the technical failure of the German scientists (led by Heisenberg) during the Second World War, he would surely have withdrawn his support from an Allied bomb-making project—as discussed in the article in this book by Joseph Rotblat, the anti-nuclear weapons campaigner who resigned from the Manhattan Project.

Einstein in his study at Princeton with Lieutenant Comstock of the US Navy, for which Einstein did research work during the Second World War, 1943.

J. Edgar Hoover feared Einstein's influence too, as did Senator Joseph McCarthy. That is why the FBI kept its Einstein investigation so secret, and why Einstein was never subpoenaed by the various Congress committees investigating alleged subversion and espionage in the 1940s and 50s. They knew that he would poke his tongue out—as he had in the famous photograph taken in Princeton on his birthday in 1951 (see page 239)—and probably make fools of them. When the public confrontation with Einstein finally came in 1953–54, it helped to turn the tide against the climate of fear and precipitate the decline of McCarthyism.

In this period Einstein made a number of public statements and supported several individuals threatened with dismissal from their jobs. But the one that really stirred public controversy was Einstein's letter to a New York teacher of English, William Frauenglass, in May 1953. When it was published in the *New York Times* with Einstein's permission, he feared that, at the age of 74 and in poor health, he might have to go to jail. Frauenglass had refused to testify before a Congress committee about his political affiliations and now faced dismissal from his school. He asked for advice from Einstein, who wrote:

The reactionary politicians have managed to instil suspicion of all intellectual efforts into the public by dangling before their eyes a danger from without... What ought the minority of intellectuals to do against this evil? Frankly, I can only see the revolutionary way of non-cooperation in the sense of Gandhi's. Every intellectual who is called before one of the committees ought to refuse to testify, i.e., he must be prepared for jail and economic ruin, in short, for the sacrifice of his personal welfare in the interest of the cultural welfare of his country... If enough people are ready to take this grave step they will be successful. If not, then the intellectuals of this country deserve nothing better than the slavery which is intended for them.

Immediately, McCarthy told the *New York Times* that "anyone who gives advice like Einstein's to Frauenglass is himself an enemy of America... That's the same advice given by every Communist lawyer that has ever appeared before our committee." A week later, he modified "enemy of America" to "a disloyal American."

No doubt many Americans agreed with this second verdict. Clearly Hoover did. Einstein, for obvious reasons, got right under Hoover's skin. But despite the best efforts of the FBI, aided by military intelligence, the CIA and other government agencies, Hoover could not convict Einstein of being a Communist, still less of helping the Soviet Union. A few days after Einstein's death, Hoover officially closed the Einstein case. At the same time, a eulogistic President Eisenhower spoke for the America that Einstein had admired: "Americans are proud that he sought and found here a climate of freedom in his search for knowledge and truth... No other man contributed so much to the vast expansion of twentieth-century knowledge."

Einstein with Elsa at the Grand Canyon, 1931.

12. Zionism, the Holocaust and Israel

"I am neither a German citizen, nor is there anything in me which can be designated as 'Jewish faith.' But I am a Jew and am glad to belong to the Jewish people, even if I do not consider them in any way God's elect."

Einstein, letter to the Central Association of German Citizens of the Jewish Faith, 1920

Einstein outside the Palestinian pavilion, New York World's Fair, 1939.

Among Einstein's many visitors in Princeton in the last year of his life was a German who wanted to interview him about religion. William Hermanns was a veteran of the First World War, a survivor of the battle of Verdun. After the war he had started a career as a diplomat but left the German foreign service with the coming of the Nazis and emigrated to the United States; there he had taught German literature at Harvard University and worked at other universities. Hermanns had first interviewed Einstein in Berlin in 1930 and then several times more; decades later, he would publish a book about him, *Einstein and the Poet: In Search of the Cosmic Man.* He shared Einstein's belief in a 'cosmic' religion, an idea which Einstein had hinted at in his conversation about reality with Rabindranath Tagore in Berlin in 1930 and which he then named and developed in "Religion and science," a controversial article published in the *New York Times* just before he arrived in the US in 1930. During the 1930s and 40s Einstein expanded upon the idea while living in the US. Hermanns felt that this cosmic religion need not negate traditional beliefs and wanted to found a cosmic-religion movement including the Jewish, Christian, Vedic, Buddhist and Islamic traditions. He was therefore keen to elicit from Einstein what he called "precise statements on God."

Concise and pithy as ever, Einstein obliged Hermanns with the following remarks on his religious beliefs:

> *About God, I cannot accept any concept based on the authority of the Church. As long as I can remember, I have resented mass indoctrination. I do not believe in the fear of life, in the fear of death, in blind faith. I cannot prove to you that there is no personal God, but if I were to speak of him, I would be a liar. I do not believe in the God of theology who rewards good and punishes evil. My God created laws that take care of that. His universe is not ruled by wishful thinking, but by immutable laws.*

Three aspects of this statement are especially notable. First, Einstein does not explain how human morality fits into his cosmic religion. Indeed he believed that, as he said elsewhere, "There is nothing divine about morality; it is a purely human affair." Secondly, he appears to leave no room in his religion for free will. For how can free will exist in a world ruled by immutable divine laws? Thirdly, he makes no reference of any kind to the human need for religious traditions, in his case Judaism.

"There is only one fault with [Einstein's] cosmical religion: he put an extra letter in the word—the letter 's'"—was the tart response from one well-known religious commentator in the US, the author of a 1920s book defending religion against science, *Religion Without God*, which Einstein had read and found to be intelligent. This was Bishop Fulton Sheen, an arch-conservative Catholic (reported to have said to his close friend, the FBI director J. Edgar Hoover: "Edgar, I'm

Einstein with Elsa at a reception in Tel Aviv, where he was made an honorary citizen of the city during his visit, 1923.

Einstein speaks at a conference of Jewish students in Germany, 1924, at a time of increasing anti-Semitism in the country.

used to pomp and ceremony, but I'm always impressed when I'm around you. Your FBI exceeds anything the Pope has ever done!"). In Sheen's view a personal God at the head of a church was a *sine qua non* of any worthwhile religion.

For Einstein, his cosmic religion and his Jewishness were apparently unrelated. Writing on the question "Is there a Jewish point of view?" he virtually said that Judaism was not a religion at all:

> *In the philosophical sense there is, in my opinion, no specifically Jewish point of view. Judaism seems to me to be concerned almost exclusively with the moral attitude in life and to life... Judaism is thus no transcendental religion; it is concerned with life as we live it and as we can, to a certain extent, grasp it, and nothing else. It seems to me, therefore, doubtful whether it can be called a religion in the accepted sense of the word, particularly as no 'faith' but the sanctification of life in a supra-personal sense is demanded of the Jew.*

He did, however, identify one strain in Judaism that was particularly highly developed, as shown in the Psalms: "a sort of intoxicated joy and amazement at the beauty and grandeur of this world, of which man can form just a faint notion." And he related this joy to "the feeling from which true scientific research draws its spiritual sustenance, but which also seems to find expression in the song of birds." (That is, the joy to which he said Marie Curie was deaf.) But, said Einstein bluntly, "To tack this feeling to the idea of God seems mere childish absurdity."

Einstein's relationship with Jewishness and Zionism was therefore "not founded on religion," Max Jammer, a physicist who knew Einstein personally and is the author of *Einstein and Religion*, explains in his article in this book.

Perhaps we can best begin to understand the complexity of the relationship from the phrase that Einstein would always use when referring to fellow Jews: not 'co-religionists' but 'tribal companions' (as in his previous comment to Fritz Haber in 1921 about his forthcoming first visit to the US with the Zionists). Einstein felt that he was bound to other Jews not by ties of religion but by tribal ties. His earliest use of the phrase appears to have been in 1914 in a letter he sent from Berlin rejecting an invitation from the Academy in St Petersburg to visit Russia, because of Tsarist Russia's history of anti-Jewish pogroms: "It goes against the grain to travel without necessity to a country where my tribal companions were so brutally persecuted."

In his early days, Einstein knew very well that his search for an academic post had been hampered by his Jewishness. A 1901 letter from his future wife Mileva Marić to her best friend makes it plain: "you know that my darling has a very wicked tongue and on top of it he is a Jew." Discreet anti-Semitism was evident in his first successful academic appointment at Zurich University in 1909. The dean of the department wrote confidentially:

> *Herr Dr Einstein is an Israelite, and...the Israelites are credited among scholars with a variety of disagreeable character traits, such as importunateness, impertinence, a shopkeeper's mind in their understanding of their academic position, etc., and in numerous cases with some justification. On the other hand, it may be said that among the Israelites, too, there are men without even a trace of these unpleasant characteristics and that it would therefore not be appropriate to disqualify a man merely because he happens to be a Jew.*

Einstein's own attitude at this time was summarized in his private comment on the Jews from wealthy families in Zurich who were *Privatdozents* (teaching assistants—the position for which he himself was rejected in 1907) but who continued to aspire to be professors, purely for reasons of social acceptance, despite being repeatedly passed over for promotion. "Why are these fellows, who make out very comfortably by private means, so anxious to land state-paid positions? Why all that humble tail-wagging to the state?" Their subservience showed a lack of proper pride, he felt.

Though his immediate family believed in a high degree of assimilation into German society, Einstein himself was less persuaded, and moved further and further away from this view with age. Haber's desire to be a Prussian—to the extent of having himself baptized a Protestant—was quite beyond the pale for Einstein. "I am delighted to know... that your former love for the blond beast has somewhat cooled off," he mocked Haber after he had to leave Germany in 1933. Einstein never felt, as Oppenheimer apparently did, that he "was Jewish, but he wished he weren't and tried to pretend that he wasn't." And he did not agree with his friend Max Born, who came from a highly assimilated family which regarded the anti-Semitic expressions and measures of pre-1918 Germany as "unjustified humiliations." Einstein's basic view, prior to the appalling excesses of the Nazi period, was that anti-Semitism, though undoubtedly unpleasant, was to be expected in any multi-ethnic society, and was "not to be got rid of by well-meaning propaganda. Nationalities want to

A rare image of Einstein wearing a kippah, during a charity concert given in 1930 by Hermann Jadlowker, star of the Berlin State Opera and the Metropolitan Opera in New York, in the great synagogue in Berlin's Cranienburgerstrasse. Einstein also performed in the concert.

pursue their own path, not to blend. A satisfactory state of affairs can only be brought about by mutual toleration and respect." Born admitted in the 1960s: "History has shown that Einstein was the more profound."

The corollary of this, for Einstein, was that Jews should build up their own sense of self-assurance and look after their own kind, rather than seeking acceptance and help from their host societies. Rabindranath Tagore took a similar line with Indians and the colonial power Britain at the same period, which is why Einstein and Tagore found much in common in their social attitudes (as noted by Isaiah Berlin), if not in their philosophies of science. And this growing feeling of solidarity with the Jewish tribe was, of course, what first sparked Einstein's interest in Zionism in 1919 and after. In Einstein's talks with the journalist Alexander Moszkowski in 1919–20, published in *Conversations with Einstein*, Judaism and Zionism are not even mentioned. It was the anti-Semitism he experienced in Germany in 1920 and after that sharpened his commitment and drew him into the Zionist fold.

Being Einstein, however—the self-confessed "lone traveller"—he never actually joined a Zionist organization. Freedom and independence always came first for him; tribal loyalty second. From 1921 onwards he would be selfless in helping the Zionists to raise money,

Einstein on his first trip to the USA, a fundraiser for the Zionist movement in 1921. Pictured here on ship are: Menachem Ussishkin (see page 189), Chaim and Vera Weizmann, Einstein and Elsa, and Benzion Mossinson.

Einstein, guided by the leading Zionist Meyer Weisgal, attends the Anglo-American Committee in Washington DC, 1946, which was considering the situation in Palestine. On the right is Helen Dukas, Einstein's secretary for the last 25 years of his life.

especially for the Hebrew University, but he would not toe their line when he disagreed with their nationalism. Chaim Weizmann knew and accepted this—deliberately keeping Einstein only as a figurehead in his 1921 speaking tour of the United States, not as a lecturer on Zionism—but it made for a prickly synergy. At one point in the 1930s the Einstein-Zionist relationship deteriorated so much that Einstein privately called Weizmann "a complete liar (a Jewish Alcibiades)," while Weizmann said that Einstein was "acquiring the psychology of a 'prima donna' who is beginning to lose her voice."

In February 1923, Einstein paid his first visit to Palestine on his way back to Europe from lecturing in Japan. He was of course fêted by his 'tribal companions' in the Holy Land and he expressed his sincere pleasure at their welcome and their pioneering constructive work. But his diary shows that the experience was also uneasy for him. At the Wailing Wall, he watched the orthodox Jews with beards and side-locks, dressed in their long black caftans and broad hats, rocking back and forth in prayer, and noted "dull-minded tribal companions... A pitiful sight of men with a past but without a future." On his last evening in Jerusalem, his hosts to a man begged him to come and live in the city and work at the Hebrew University. He noted: "My heart says yes, but my reason says no!" He never returned to the Holy Land.

The strains nearly wrecked his relationship with the university. In 1933 he wrote savagely to Born (who was looking for a post after leaving Nazi Germany) that the Jerusalem University had become "a nasty mess... complete charlatanism." At the root of the conflict was Einstein's unwillingness to compromise on academic standards. He would rather have no university than a mediocre one. The president of the university, Judah Magnes, was utterly unacceptable to him. An American rabbi, Magnes had been chosen by the university's Jewish sponsors, who were more interested in arranging positions for people from wealthy American Jewish families than in scholarship. Einstein thought Magnes an "ambitious and weak person [who] surrounded himself with other morally inferior men, who did not allow any decent person to succeed there." Einstein therefore resigned from the university in 1928; but only after he went public with his dissatisfaction in 1933, to the irritation of Weizmann, was a commission of inquiry finally appointed. From 1935 onwards Magnes ceased to have any control over academic appointments and a rector whom Einstein admired was appointed. His relationship now mended. Before his death he willed that the last depository of all his papers should be the Hebrew University.

Helping Jewish refugee physicists like Born find jobs outside Germany was a major preoccupation for Einstein after 1933. It was among the reasons he wanted to meet President Roosevelt when he arrived in the US. As the Jewish persecution by the Nazis worsened, Einstein tirelessly lobbied for numerous refugees, including physicists of note, and found himself at frequent loggerheads with the Roosevelt administration, which was by no means anti-Nazi in the mid to late 1930s. Relations between the FBI and the Gestapo remained close right up to the Japanese attack on Pearl Harbor in 1941. "If the Nazis claimed you had 'Communist sympathies,' that was all the United States authorities needed—you were barred," writes Fred Jerome in *The Einstein File*. Einstein could not place even his visiting foreign assistants in permanent posts at the Institute for Advanced Study, since he had fallen out with Flexner, the director, in 1933. But at least in one unusual case this perilous situation led to a happy solution. Einstein's Polish collaborator Leopold Infeld suggested that the two of them write a book together. *The Evolution of Physics*, which appeared under their joint authorship in 1938, was a bestseller, such was the magic of Einstein's name (and a fascinating work, much quoted in this book); its commercial success may also have saved its co-author Infeld from incineration by the Nazis.

When the news of the death camps began to leak out towards the end of the war, Einstein's attitude to the Germans became implacable. In 1944 he sent a message to a bulletin run by Polish Jews in New York, entitled "To the heroes of the battle of the Warsaw Ghetto"—referring to the desperate Jewish uprising, massacre and subsequent extermination in Poland the previous year. He wrote:

The Germans as an entire people are responsible for these mass murders and must be punished as a people if there is justice in the world and if the consciousness of collective responsibility in the nations is not to perish from the earth entirely. Behind the Nazi party stands the German people, who elected Hitler after he had in his book and in his speeches made his shameful intentions clear beyond the possibility of misunderstanding. The Germans are the only people who have not made any serious attempt of counteraction leading to the protection of the innocently persecuted. When they are entirely defeated and begin to lament over their fate, we must not let ourselves be deceived again, but keep in mind that they deliberately used the humanity of others to make preparation for their last and most grievous crime against humanity.

After the war, Einstein withdrew permission for new editions of his works in Germany and rebuffed all official German attempts to honour him, even purely scientific ones, without exception. When Otto Hahn, the discoverer of nuclear fission in 1938, invited Einstein to become a 'foreign scientific member' of the newly formed Max Planck Society in 1949, Einstein informed him that, "The attitude of the German intellectuals—viewed as a class—was no better than that of a mob. There is not even remorse or an honest desire to make good whatever, after the gigantic murdering, is left to be made good." He even admonished Born when he announced his intention of retiring from Edinburgh to live in Germany again, for, as Einstein put it, "migrating back to the land of the mass-murderers." But he remained on friendly terms with a few of his old German colleagues as individuals, especially with Max von Laue, who had been the only scientist in the Prussian Academy with the courage to protest against its official abuse of Einstein after his resignation in 1933. When Planck died in 1947, Einstein wrote to his widow: "The hours which I was permitted to spend at your house, and the many conversations which I conducted face to face with that wonderful man, will

remain among my most beautiful recollections for the rest of my life. This cannot be altered by the fact that a tragic fate tore us apart."

Einstein resisted, too, being drawn into official activities by the new state of Israel—though obviously for quite different reasons. In the 1930s he had opposed the formation of a Jewish state on the grounds that it would damage "the essential nature of Judaism" and had constantly encouraged cooperation between Jews and Arabs. But after the war and the Holocaust he accepted Israel as a *fait accompli.* Nevertheless he could not change the anti-political habits of a lifetime. When Weizmann, the first president of Israel, died in November 1952 and the Israeli prime minister David Ben-Gurion offered the presidency to Einstein, he immediately declined.

"I am deeply moved by the offer from our state Israel, and at once saddened and ashamed that I cannot accept it. All my life I have dealt with objective matters, hence I lack both the natural aptitude and the experience to deal properly with people and to exercise official functions," Einstein explained. "I am the more distressed over these circumstances because my relationship with the Jewish people has become my strongest human bond, ever since I became fully aware of our precarious situation among the nations of the world."

Ben-Gurion must have been moved, but relieved also, by Einstein's response. While awaiting his decision the prime minister told his assistant, "Tell me what to do if he says yes! I've had to offer the post to him because it's impossible not to. But if he accepts we're in for trouble." Ben-Gurion well knew that Einstein could never be naturally comfortable with authority, whether it was German, American or Israeli.

Einstein at home in Princeton with a group of Jewish refugee children, 1949.

Einstein on Religion, Judaism and Zionism

Max Jammer

Next to physics, the philosophy of religion and the quest for spiritual truth was perhaps the chief preoccupation of Albert Einstein—so much so, in fact, that he has been called a "disguised theologian." But although Einstein was deeply religious, he never attended a religious service and called himself a "deeply religious nonbeliever." His attitude to religion and Zionism is one more paradox among the many paradoxes in the physics, philosophy and social and political beliefs of a profound human being.

Max Jammer and Einstein; a still from a film shot in Princeton, 1952.

Einstein's conception of religion can be understood only if we trace its development from his early youth. At the age of six Albert, though registered as a Jew, entered a Catholic public primary school in Munich, which was less expensive than a distant private Jewish school. There he learned the catechism and read biblical tales from both the Old and New Testaments. Although his parents were irreligious Jews they hired a distant relative to teach Albert the principles of Judaism. This teacher aroused in him sentiments so strong that he observed religious prescriptions in every detail and even chided his parents for not observing Jewish dietary laws. Still, the boy saw in Judaism and Christianity only one harmonious body of doctrine.

At the age of twelve, however, just when he should have been preparing himself for the bar mitzvah, he lost what he later called his "religious paradise of youth." Ironically his conversion into a "fanatical freethinker" was caused by the only Jewish religious custom which his parents observed: to host a poor Jewish student for a weekly meal. This guest, although ten years older than Albert, became his intimate friend and gave him scientific books to read which completely shattered his religious belief.

What particularly shocked him, and led to his lifelong rejection of anthropomorphic conceptions of God, was a quotation from Xenophanes: "If oxen could paint, they would present their gods in the form of oxen." It created "a crushing impression," coupled with a feeling that "youth is intentionally being deceived by the state through lies." From now on he distrusted any kind of authority and developed a critical attitude not only to the laws of human society but also to the laws of nature, which, according to some historians of modern physics, explains his readiness to accept the revolutionary character of his theory of relativity.

A change in his anti-religious position occurred when he was working at the Swiss Patent Office in Bern and meeting with friends to discuss famous philosophers. Spinoza's philosophy made him again deeply religious—but now he conceived God rather like Spinoza's *deus sive natura* as a superior intelligence which reveals itself in the harmony and beauty of nature. He also agreed with the greatest Jewish philosopher of the Middle Ages, Moses Maimonides, who denied God any corporeality and proposed the so-called 'negative theology' which uses the method of double negations, for example 'God is not non-existent'—God's existence can be inferred only from His 'ways' or actions manifested in nature. Einstein contended that "everyone who is seriously involved in the pursuit of science becomes convinced that a spirit is manifest in the laws of the universe—a spirit vastly superior to that of man and one in the face of which we with our modest powers must feel humble."

Einstein called his religious conviction a "cosmic religious feeling" and claimed that it is "the strongest and noblest motive for scientific research," as exemplified by Kepler and Newton. Therefore science and religion are connected so intimately that "science without religion is lame, religion without science is blind."

In accordance with his credo, he never attended service in a synagogue. He even declared that the person who prays to God and asks him for something is not a religious person. Yet, as his close friend the Jewish physicist Max Born once remarked, "Einstein did not think that religious faith was a sign of stupidity, nor unbelief a sign of intelligence." In fact, Einstein repeatedly criticized atheists and wondered how "in view of such a harmony in the cosmos there are yet people who say that there is no God." Even in the harmony of music Einstein felt something divine. When he heard the young violinist Yehudi Menuhin play Bach, Beethoven and Brahms with the Berlin Philharmonic Orchestra, at the end he rushed over to Menuhin, embraced him, and exclaimed: "Now I know there is a God in heaven!" Clearly, Einstein was 'religious,' but if 'religious belief' demands membership of a religious community, he undoubtedly was an unbeliever.

His relationship with Zionism, which we now turn to, was therefore not founded on religion. During his Swiss years he was uninvolved in Jewish affairs. Not until 1911, when he was offered a full professorship in Prague and was obliged to declare his religious affiliation, did Einstein call himself a Jew. It was also in Prague that for the first time he came into daily contact with Jews as a community, which formed half of the German-speaking population of the city. But attempts to involve him in Jewish activities by the novelist Max Brod and the philosopher Hugo Bergmann failed, for Einstein was utterly absorbed in physics.

"On a Jewish Palestine": first version of a speech given by Einstein to a Zionist assembly, 27 June 1921. He expresses his hope that "a homeland for our national culture" would be created in Palestine for the benefit of all its inhabitants.

Prof. Dr. A. Einstein
Mitgl. d. pr. Akademie
der Wissenschaften

Berlin, den 27. VI. 21.
W. 30, Haberlandstr. 5.

Meine Damen und Herren!

Seit zweitausend Jahren bestand das gemeinsame Gut des jüdischen Volkes nur in seiner Vergangenheit. Gemeinsam war unserem über die Welt zerstreuten Volke nichts als die sorgsam gehütete Tradition. Wohl haben einzelne Juden grosse Kulturwerte geschaffen, aber das jüdische Volk als Ganzes schien nicht mehr die Kraft zu grossen Kollektiv-Leistungen zu haben.

Dies ist nun anders geworden. Die Geschichte hat uns eine grosse und edle Aufgabe zugewiesen in Gestalt der thätigen Mitarbeit an dem Aufbau Palästinas. Hervorragende Stammesgenossen arbeiten bereits mit allen Kräften an der Verwirklichung dieses Zieles. Es ist uns Gelegenheit dazu geboten Kulturstätten zu errichten, die das ganze jüdische Volk als sein Werk betrachten kann. Wir hegen die Hoffnung in Palästina eine Heimstätte eigener nationaler Kultur zu schaffen, die dazu beitragen soll, den nahen Orient zu neuem wirtschaftlichen und geistigen Leben zu wecken.

Das Ziel, das den Führern des Zionismus vorschwebt, ist kein politisches, sondern ein sociales u. kulturelles. Das Gemeinwesen in Palästina soll sich dem socialen Ideal unserer Vorfahren nähern, so wie es in der Bibel niedergelegt ist und gleichzeitig eine Stätte modernen geistigen Lebens werden. Ein geistiges Centrum für die Juden der ganzen Welt. Dieser Auffassung entsprechend bildet die Errichtung einer jüdischen Universität in Jerusalem eines der wichtigsten Ziele der zionistischen Organisation. Ich bin in den letzten Monaten in Amerika gewesen, um dort die materielle Basis für diese Universität schaffen zu helfen. Der Erfolg dieser Bestrebung war ein vorzüglicher. Dank der unermüdlichen Tätigkeit und der hervorragenden Opferwilligkeit der jüdischen Ärzte Amerikas ist es uns gelungen, genügend Mittel für die Schaffung einer medizinischen Fakultät zusammen zu bringen u. es wird mit den vorbereitenden Arbeiten derselben sofort begonnen. Nach den bisherigen Erfolgen hege ich keinen Zweifel, dass sich die materielle Basis für die übrigen Fakultäten in kurzer Zeit wird schaffen lassen. Die med. Fakultät soll zunächst im Wesentlichen als Forschungsinstitut ausgebildet werden u. für die Aufbau besonders wichtige Sanierung des Landes tätig sein. Unterrichtstätigkeit in grösserem Stile wird erst später von Wichtigkeit werden. Da sich eine Reihe tüchtiger Forscher bereits gefunden hat, die

It was only after he moved to Berlin in 1914, divorced his Catholic wife Mileva and married his cousin Elsa, that Jews attracted his attention, for Berlin had at that time a strong influx of Jews, seeking refuge from cruel persecutions in Poland and Russia. Realizing that wartime demagogues were using the influx as a political weapon Einstein tried, merely for humanitarian reasons, to help these unfortunate but often talented people. Since they were debarred from attending the university he arranged special academic courses for them, often taught by himself.

The year 1919 was of course momentous for Einstein. He became world famous with the announcement of the astronomical observations that verified the predictions of the general theory of relativity. He remarked jokingly in the London *Times*: "By an application of the theory of relativity to the taste of readers, today in Germany I am called a German man of science, and in England I am represented as a Swiss Jew. If I come to be regarded as a *bête noire*, the descriptions will be reversed, and I shall become a Swiss Jew for the Germans and a German man of science for the English!"

In addition, 1919 saw the beginning of Einstein's conversion to Zionism. In February Kurt Blumenfeld, a Zionist leader in Berlin, tried to enlist him in the Zionist Organization by making reference to the tragic situation of Jews in Europe. Einstein replied: "What has that to do with Zionism?... Are not the Jews, through a religious tradition which has evolved outside Palestine, too much estranged from the country and country life?" But he was sympathetic. Although Blumenfeld failed to enlist Einstein officially, he succeeded in enkindling his Jewishness. In early 1920, when the Central Association of German Citizens of the Jewish Faith invited him to a meeting, Einstein excused himself: "I am neither a German citizen, nor is there anything in me which can be designated as 'Jewish faith.'" Yet he also stated: "But I am a Jew and am glad to belong to the Jewish people, even if I do not consider them in any way God's elect." Indeed, less than a month before his death in 1955, Einstein wrote to Blumenfeld: "I thank you, even at this late hour, for having made me conscious of the Jewish soul."

Shortly afterwards, in 1921, Chaim Weizmann, then president of the Zionist Organization, asked Einstein to join him on a fund-raising tour of the United States for the establishment of a university in Jerusalem. This time Einstein agreed. "It was in America," Einstein said in a speech after his return to Berlin, "that I first discovered a Jewish people. I have seen any number of Jews, but a

Jewish people I have never met either in Berlin or elsewhere in Germany. This Jewish people, coming from Russia, Poland, or other East European countries...still retain a healthy national feeling, not yet destroyed by atomization and dispersion. I found these people extraordinarily ready for self-sacrifice and practically creative."

Einstein plants a tree, near Haifa, on his 1923 visit to Palestine.

His sole visit to Palestine took place in 1923. During his twelve days there Einstein delivered the first lecture of the Hebrew University on Mount Scopus. Menachem Ussishkin, president of the Zionist Executive, who invited him to settle in Jerusalem, introduced him with these words: "Mount the platform which has been waiting for you for 2000 years." Einstein began with a sentence in Hebrew, continued in French and concluded in German. The following day, at a banquet held in his honour in Tel Aviv, he was declared a free citizen, to which he replied: "I have already had the privilege of receiving the honorary citizenship of the City of New York, but I am tenfold happier to be a citizen of this beautiful Jewish town." In 1952, Einstein could even have become the head of the Jewish state, when Prime Minister Ben-Gurion offered him the presidency of Israel in order to demonstrate that culture and science are the country's highest ideals. But Einstein declined, saying "I know a little about nature, and hardly anything about men."

He became an active member of the first board of governors and chairman of the academic council of the Hebrew University, which he called "the greatest thing in Palestine since the destruction of the Temple of Jerusalem," and which he expected to become a centre of intellectual life not only for Jews. But soon Einstein disagreed with the policy of the university's chancellor, Judah Magnes, who in his view was over-influenced by American philanthropists in handling professional affairs instead of leaving them to the academic council of the university. In 1928 Einstein resigned his official positions but wrote that he would "never cease to regard the fate of the Jerusalem university as a matter close to my heart." When in 1935 the university adopted his view he revoked his resignation and continued to support its growth into what he hoped would become "a great spiritual centre which will evoke the respect of cultured mankind the world over." Einstein's will bequeathed his literary rights and his personal archives, with his manuscripts and correspondence, to the Hebrew University.

The Jewish-Arab conflict worried Einstein very much. After the 1929 riots in Hebron, in which many Jews were murdered by Arabs, he rejected any violent retaliation. He was convinced that no irreconcilable differences stood in the way of peace between Jews and Arabs. He wrote to Weizmann: "Should we be unable to find a way to honest cooperation and honest pacts with the Arabs, then we shall have learnt nothing from our 2000 years of suffering and will deserve our fate." In a letter to the publisher of the Arab daily *Falastin* in 1930 he put forward the idea of a privy council, consisting of four Jews and four Arabs, meeting once a week for deliberations in order to achieve the maximal "welfare of the whole population of the country."

Half a century after Einstein's death, we may well ask what he might have to say about the present political situation in Israel. Many of his hopes for the future of the Jewish people in its homeland have become reality. Industry, agriculture, commerce, technology and science have made progress, perhaps more than he could have anticipated. But one item, closest to his heart, has not yet been achieved: peace.

To the Arabs Einstein would most probably repeat what he wrote in another letter to *Falastin* around the same time: "One who, like myself, has cherished for many years the conviction that the humanity of the future must be built up on an intimate community of the nations, and that aggressive nationalism must be conquered, can see a future for Palestine only on the basis of peaceful cooperation between the two peoples who are at home in the country. For this reason I should have expected that the great Arab people will show a truer appreciation of the need which the Jews feel to rebuild their national home in the ancient seat of Judaism."

To the Jews he would probably repeat what he said in a broadcast for the *United Jewish Appeal* in 1949: "When appraising the achievement,...let us not lose sight of the cause to be served by this achievement:...creation of a community which conforms as closely as possible to the ethical ideals of our people as they have been formed in the course of a long history. One of these ideals is peace, based on understanding and self-restraint, and not on violence...we *want* peace and we realize that our future development depends on peace."

13. Nuclear Saint and Demon

"The feeling for what ought and ought not to be grows and dies like a tree, and no fertilizer of any kind will do very much good. What the individual can do is to give a fine example, and to have the courage to uphold ethical convictions sternly in a society of cynics. I have for a long time tried to conduct myself in this way, with a varying degree of success." Einstein, letter to Max Born, 1944

A year after the testing of the world's first atomic bomb, the cover of *Time* magazine for 1 July 1946 carried a striking painting of "Cosmoclast Einstein." In the foreground is the wild-maned professor's benevolent face, aged and weary, his gaze directed far away, apparently on posterity, while in the background a tall pillar of smoke and flame rises heavenwards into a mushroom cloud rearing like a cobra's hood over the intervening ocean—where tiny warships, dwarfed, float defencelessly. Written in the cloud itself, like some evil genie, is the equation $E=mc^2$. The implication seemed to be that Einstein's genius had conjured up this fundamental mathematical formula, applied it in a Faustian pact with the laws of theoretical physics in order to make an atomic bomb, and was now meditating upon its sin of bringing such a monster into the world of human beings.

In reality this was quite far from the truth. Einstein did not give birth to the atomic bomb. He was certainly

After the USA dropped atomic bombs on Hiroshima and Nagasaki in 1945, Einstein became associated, through his most famous equation, with the horrors of the atomic age, as this cover of *Time* magazine from 1946 makes clear.

responsible for discovering $E=mc^2$, which he published in 1905; but when he derived the equation from his theory of special relativity he had absolutely no vision of its use in weaponry. And while he did encourage the United States government to develop an atomic bomb out of his deep and reasonable fear that the Nazis would make one first, he made no contribution to understanding the physics and chemistry of atomic fission and played no role at all in the wartime Manhattan Project. On the other hand, he did firmly regret his earlier encouragement of the government once he came to know that the German scientists had had no chance of succeeding; and for the remainder of his life he was relentlessly opposed to the spread of nuclear weapons. Thus, although Einstein was definitely not the 'father' of the atomic bomb implied by *Time* magazine—as Joseph Rotblat reminds us in his article—there is a case to be made for calling Einstein the bomb's 'grandfather.'

Otto Hahn (left), who worked with Fritz Strassmann (opposite) and discovered nuclear fission in 1938. Lise Meitner (right), Hahn's former collaborator but by then a refugee from Germany, explained their experimental results theoretically.

Perhaps surprisingly, the possibilities for the massive energy locked up inside the atom had been predicted even before the publication of Einstein's equation. The predictions arose out of experimental work on radioactivity, not out of relativity theory. As early as 1904, the physical chemist Frederick Soddy (who won the Nobel prize in chemistry for the same year as Einstein, 1921) told the Corps of Royal Engineers in a lecture:

> *It is possible that all heavy matter possesses, latent and bound up in the structure of the atom, a similar quantity of energy to that possessed by radium. If it could be tapped and controlled, what an agent it would be in shaping the world's destiny. The man who put his hand on the lever by which a parsimonious nature regulates so jealously the output of this store of energy would possess a weapon with which he could destroy the Earth if he chose.*

Five years later, in his popular book *The Interpretation of Radium*, Soddy struck an optimistic note instead (though also disturbing, unintentionally): "The nation which can transmute matter could transform a desert continent, thaw the frozen poles, and make the whole world one smiling garden of Eden."

By the time that *Conversations with Einstein* appeared in late 1920, such speculations about atomic

energy (though not yet its link with $E=mc^2$) were becoming common. But Einstein was unconvinced. He told the book's author Alexander Moszkowski:

> *At present there is not the slightest indication of when this energy will be obtainable, or whether it will be obtainable at all. For it would presuppose a disintegration of the atom effected at will—a shattering of the atom. And up to the present there is scarcely a sign that this will be possible. We observe atomic disintegration only where nature herself presents it, as in the case of radium...*

Then, during the course of their conversations, Ernest Rutherford announced that he had achieved such a shattering of the nucleus, by bombarding nitrogen with alpha particles (which transmuted it into oxygen). Einstein now told Moszkowski that Rutherford's experiments had modified his view and that the verdict on atomic energy would have to await further scientific developments in nuclear distintegration.

Yet in 1933 Rutherford still thought that anyone who talked of nuclear power was "talking moonshine," as he told the London *Times*. Einstein remained unconvinced too. In a lecture in late 1934 at the annual meeting of the American Association for the Advancement of Science, he told eager reporters with his usual wit that the chance of releasing atomic energy by bombarding the atomic nucleus was about as likely as hitting birds by firing guns at them in the dark "in a neighbourhood that has few birds." And on his sixtieth birthday in March 1939, three months after the discovery in Nazi Germany of nuclear fission in uranium by Otto Hahn and Fritz Strassmann, he nevertheless insisted: "The results gained thus far concerning the splitting of the atom do not justify the assumption that the atomic energy released in the process could be economically utilized." However he added significantly: "there can hardly be a physicist with so little intellectual curiosity that his interest in this important subject could become impaired because of the unfavourable conclusion to be drawn from past experimentation."

The problem, for Einstein, was that in nuclear physics he was not ahead of the game, because in the 1930s he had progressively lost interest in most physics except physics which he judged relevant to the search for a unified theory. When Niels Bohr came to the Institute for Advanced Study in Princeton in early 1939 to discuss the latest ideas on atomic fission, Einstein remained totally detached.

Four months later, in July, he awoke to the latest possibilities. The trigger was a visit from two Hungarian émigré physicists anxious about Hitler, Leo Szilard and Eugene Wigner—the first of whom was a friend. (Amongst their common interests was the refrigerator Einstein had designed with Szilard in the 1920s.) So concerned were these two by reports that German physicists were investigating fission, very likely with a view to making a bomb, that they decided to drive out from New York and interrupt Einstein at his summer house in rural Long Island, in order to ask him to intervene politically. It was so tricky to find the address in Peconic, however, that they were about to give up and return to New York when Szilard thought of asking a young boy where Professor Einstein lived. The boy got into their car and took them to him.

Talking to Einstein, Szilard was surprised to discover that he had not considered the possibility of a nuclear chain reaction. That is, the idea of one neutron bombarding a uranium atom, causing it to fission and

SECOND NEWS SECTION

For News, Subscriptions or Advertising Call ATlantic 6100.

SATURDAY MORNING,

Pittsburgh Post-Gazette

DECEMBER 29, 1934.

SPORTS, FINANCIAL, CLASSIFIED SECTION

This WORLD of OURS

FOR MEN ONLY

Constantly Expressive—Einstein Warms Up to His Subject

ATOM ENERGY HOPE IS SPIKED BY EINSTEIN

Efforts at Loosing Vast Force Is Called Fruitless.

SAVANT TALKS HERE

Now Indicates Doubt Of Relativity Theory He Made Famous.

Blind chance, or cause and effect—which ever you prefer—may run the universe.

Space may be infinite, or it may be finite, nobody knows.

It may be curved or not curved, just as you please.

Whatever you decide, no one can contradict you, because no one has so far been able to prove any one of these contentions. But, you may be wrong, because some of the contentions may be proved in the future.

That is the contention of Prof. Albert Einstein.

But the "energy of the atom" is something else again. If you believe that man will someday be able to harness this boundless energy—to drive a great steamship across the ocean on a pint of water, for instance—then, according to Einstein, you are wrong now.

Energy of Atom.

The idea that man might some day utilize the atom's energy brought the only emphatic denial from the noted scientist yesterday when he was interviewed by a score of newspapermen at the home of Nathaniel Spear, near

A 1934 newspaper headline, "Atom energy hope is spiked by Einstein," underlines how for many years Einstein never believed it would be possible for scientists to harness the awesome energy locked up in the atom, for good or for ill.

release two neutrons, which then cause two uranium atoms to fission, producing four neutrons, and so on—and very quickly a concatenation of neutrons and an explosion of atomic energy. "I never thought of that!" Szilard recalled Einstein's saying when he told him that Enrico Fermi (a recent refugee from Mussolini's Italy) had just achieved a nuclear chain reaction in his New York laboratory. But Einstein as usual was quick to see the implications of the new idea. "He was willing to assume responsibility for sounding the alarm even though it was quite possible that the alarm might prove to be a false alarm," said Szilard. "The one thing most scientists are really afraid of is to make fools of themselves. Einstein was free from such a fear and this above all is what made his position unique on this occasion."

By 2 August, they had the historic letter from Einstein to President Roosevelt ready. Its most dramatic paragraph read as follows:

This new phenomenon [a nuclear chain reaction] would also lead to the construction of bombs, and it is conceivable—though much less certain—that extremely powerful bombs of a new type may thus be constructed. A single bomb of this type, carried by boat and exploded in a port, might very well destroy the whole port together with some of the surrounding territory. However, such bombs might very well prove to be too heavy for transportation by air.

The letter was personally delivered to the president through a trusted intermediary after a considerable delay—by which time war had broken out in Europe. Roosevelt responded promptly, but it took well over two years (and another reminder to Roosevelt from Einstein in 1940), and the Japanese attack on the United States in December 1941, before the Manhattan Project got fully underway.

The first atomic bomb was dropped on Hiroshima on 6 August 1945. Einstein's secretary Helen Dukas

Einstein with J. Robert Oppenheimer, Princeton, late 1940s. Oppenheimer was director of the Institute for Advanced Study.

heard the news on the radio and passed it on. "*Oj weh* (Woe is me)" was Einstein's only reaction. He must surely have thought of the city he had seen on his romantic visit to Japan in the early 1920s. A year or so later, he bought a thousand copies of the journalist John Hersey's horrendous and moving account of the American bombing, *Hiroshima* (first published in the *New Yorker*), and distributed them to friends.

Very soon after the end of the war Einstein began his public campaign to control nuclear weapons by calling for a new political ethics, which culminated in his televised appeal in 1950 against the development of the hydrogen bomb in the programme hosted by Eleanor Roosevelt. In 1945 he hoped that the fresh horrors of the Second World War and the obvious potential horrors of a nuclear Third World War might together be enough to force reform in international affairs. "It is consistent with what we know about Einstein's way of thinking that he believed a fundamental reform of concepts to be as necessary in ethics as in physics," writes Freeman Dyson in his book *Imagined Worlds*.

Einstein's speech in December 1945 on the occasion of the fifth Nobel anniversary dinner in New York can stand for his many public statements on the subject of this new ethics. In the speech he coined a phrase which for many captured the post-war disappointment with existing politics: "The war is won, but the peace is not."

He began: "Physicists find themselves in a position not unlike that of Alfred Nobel. He invented the most powerful explosive ever known up to his time, a means of destruction par excellence. In order to atone for this, in order to relieve his human conscience, he instituted his awards for the promotion of peace and for achievements of peace."

Then Einstein went on to describe, in searing terms, how little had been done by the Allied nations to help defenceless Jews because it was not in Allied interests to do so—either before the war, during the war or now, after the end of the conflict. He admonished his audience: "The fact that many of them are still kept in the degrading conditions of concentration camps by the Allies gives sufficient evidence of the shamefulness and hopelessness of the situation." Jewish emigration to Palestine had been blocked by power politics. "It is sheer irony when the British foreign minister tells the poor lot of European Jews they should remain in Europe because their genius is needed there, and, on the other hand, advises them not to try to get to the head of the queue lest they might incur new hatred and persecution," said Einstein. "Well, I am afraid they cannot help it; with their six million dead they have been pushed to the head of the queue, of the queue of Nazi victims, much against their will."

He concluded grimly:

So far as we, the physicists, are concerned, we are no politicians and it has never been our wish to meddle in politics. But we know a few things that the politicians do not know. And we feel the duty to speak up and to remind those responsible that there is no escape into easy comforts, there is no distance ahead for proceeding little by little and delaying the necessary changes into an indefinite future, there is no time left for petty bargaining. The situation calls for a courageous effort, for a radical change in our whole attitude, in the entire political concept. May the spirit that prompted Alfred Nobel to create his great institution, the spirit of trust and confidence, of generosity and brotherhood among men, prevail in the minds of those upon whose decisions our destiny rests. Otherwise, human civilization will be doomed.

Einstein's main practical recommendation for managing nuclear weapons (which he sensibly anticipated the Soviet Union would soon develop) was that they could be controlled only by what he called a "world government." This would be an essentially military organization, to which the world's leading nations would contribute armed forces, which would then be "commingled and distributed as were the regiments of the former Austro-Hungarian Empire," and which would have the power to enforce international law according to the direction of its representative executive. "Do I fear the tyranny of a world government? Of course I do. But I fear still more the coming of another war or wars." The United States, he said, should imme-

Einstein at his summer house in Caputh, Germany, around 1930.

Albert Einstein
Old Grove Rd.
Nassau Point
Peconic, Long Island

August 2nd, 1939

F.D. Roosevelt,
President of the United States,
White House
Washington, D.C.

Sir:

Some recent work by E.Fermi and L. Szilard, which has been communicated to me in manuscript, leads me to expect that the element uranium may be turned into a new and important source of energy in the immediate future. Certain aspects of the situation which has arisen seem to call for watchfulness and, if necessary, quick action on the part of the Administration. I believe therefore that it is my duty to bring to your attention the following facts and recommendations:

In the course of the last four months it has been made probable - through the work of Joliot in France as well as Fermi and Szilard in America - that it may become possible to set up a nuclear chain reaction in a large mass of uranium,by which vast amounts of power and large quantities of new radium-like elements would be generated. Now it appears almost certain that this could be achieved in the immediate future.

This new phenomenon would also lead to the construction of bombs, and it is conceivable - though much less certain - that extremely powerful bombs of a new type may thus be constructed. A single bomb of this type, carried by boat and exploded in a port, might very well destroy the whole port together with some of the surrounding territory. However, such bombs might very well prove to be too heavy for transportation by air.

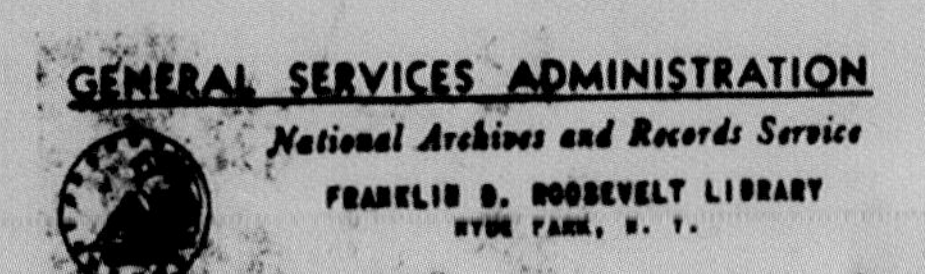

Einstein's letter to President Roosevelt, 1939, encouraging the US government to pursue research into atomic weapons.

-2-

The United States has only very poor ores of uranium in moderate quantities. There is some good ore in Canada and the former Czechoslovakia, while the most important source of uranium is Belgian Congo.

In view of this situation you may think it desirable to have some permanent contact maintained between the Administration and the group of physicists working on chain reactions in America. One possible way of achieving this might be for you to entrust with this task a person who has your confidence and who could perhaps serve in an inofficial capacity. His task might comprise the following:

a) to approach Government Departments, keep them informed of the further development, and put forward recommendations for Government action, giving particular attention to the problem of securing a supply of uranium ore for the United States;

b) to speed up the experimental work, which is at present being carried on within the limits of the budgets of University laboratories, by providing funds, if such funds be required, through his contacts with private persons who are willing to make contributions for this cause, and perhaps also by obtaining the co-operation of industrial laboratories which have the necessary equipment.

I understand that Germany has actually stopped the sale of uranium from the Czechoslovakian mines which she has taken over. That she should have taken such early action might perhaps be understood on the ground that the son of the German Under-Secretary of State, von Weizsäcker, is attached to the Kaiser-Wilhelm-Institut in Berlin where some of the American work on uranium is now being repeated.

Yours very truly,
A. Einstein
(Albert Einstein)

diately announce its readiness to commit the secret of the atomic bomb to this world government. And the Soviet Union should be sincerely invited to join it. In September 1947 Einstein proposed his idea in an open letter to the General Assembly of the United Nations. If the UN were to have a chance of becoming such a world government, he said, then "the authority of the General Assembly must be increased so that the Security Council as well as all other bodies of the UN will be subordinated to it."

Perhaps needless to say, as the Cold War hotted up in 1947, no leading power was remotely interested in Einstein's world government. It was assailed from all sides. Four top scientists of the Russian Academy replied respectfully but in unequivocal opposition. "By an irony of fate, Einstein has virtually become a supporter of the schemes and ambitions of the bitterest foes of peace and international cooperation... Essentially this proposal differs very little from the suggestions of frank advocates of American imperialism." Einstein was not surprised by the Soviet attitude but pleaded in his response that, "If we hold fast to the concept and practice of unlimited sovereignty of nations it only means that each country reserves the right for itself of pursuing its objectives through warlike means... I advocate world government because I am convinced that there is no other possible way of eliminating the most terrible danger in which man has ever found himself."

'World government' enjoyed a brief vogue among a spectrum of American intellectuals—on the right as well as on the left—and then faded from view. So too did the Emergency Committee of Atomic Scientists, a brainchild of Szilard (who worked on the Manhattan Project), which Einstein agreed to chair in May 1946. Both fell victim to the Cold War. The FBI file on Einstein was initially very concerned that the committee would leak the secrets of the atomic bomb to the Russians, until the FBI was discreetly reminded by the Atomic Energy Commission that Einstein had not been given clearance by the Army or the FBI to work on the bomb. As for world government, it was of no interest to the FBI, since the agency knew that the Russians were opposed to it.

It would be too easy to call Einstein 'naive' in his call for a new political ethics—a common charge against him. Rotblat, who founded the influential Pugwash movement against nuclear proliferation during the Cold War, does not think so. Einstein inspired him. Nor does the physicist and writer Jeremy Bernstein who calls Einstein's political thinking "tough, lucid, and 'infinitely far-seeing'... Einstein was in no way the confused idealist he is sometimes made out to be." Nor does the Harvard University physicist and historian Gerald Holton, quoted by *Time* magazine in its profile of Einstein as "Person of the Century" in 1999: "If Einstein's ideas are really naive, the world is really in pretty bad shape." Holton thinks that Einstein's vision of an ethical politics offers "an ideal political model for the twenty-first century." But it does have to be said that in his political proposals Einstein always underestimated the human need for belonging, because he himself did not rate it highly. Which is strange and ironic, given the phrase he chose for his fellow Jews—'tribal companions.' A 'world government'—or for that matter the actually existing United Nations—is never likely to inspire as much loyalty in people as national institutions—unless you are an Einstein.

In political matters, Einstein recognized only too clearly that intellectuals, even a Nobel laureate like the Nazi Philipp Lenard, can behave just as badly as politicians and leaders. He told Born in 1944: "We really should not be surprised that scientists (the vast majority of them) are no exception to this rule, and *if* they are different it is not due to their reasoning powers but to their personal stature, as in the case of Laue"—his friend in the Prussian Academy who survived the Nazi regime without compromise. "It was interesting to see the way in which [Laue] cut himself off, step by step, from the traditions of the herd, under the influence of a strong sense of justice." Thus Einstein, paradoxical as ever, in the end accepted that a human being's stature is defined not by the divine laws he sought in his physics but by his or her free moral choices.

Einstein in his study at Princeton, 1950. In the annotation to a friend ("the clever Lina"), he sends greetings from his "shoemaker's workshop."

Der pfiffigen Lina diesen Gruss
aus meiner Schusterei
A. Einstein 1950.

Einstein's Quest for Global Peace

Joseph Rotblat

Einstein's fame and unique status in the world are mainly due to his scientific discoveries. Much less is known about his political activities: his anti-war campaigns and his advocacy of a world government. Yet next to science, these matters were nearest to him; he devoted to them much time until the very end; and, as in his scientific theories, Einstein was iconoclastic in political pronouncements. It was for this reason that he occasionally came into conflict with 'official' policies.

One such occasion was when he advocated military preparedness against Hitler's Germany. Einstein recognized as early as 1933 the grave threat to democracy posed by the fast-growing military strength of the Nazi regime. His sober, realistic analysis of events led him to the conclusion that the only response was to build up a strong military force, capable of resisting a Nazi onslaught. Although the motivation was quite clear and, if it had been acted upon, might have prevented the Holocaust, the official pacifist leadership dogmatically stuck to its rigid principles and rejected Einstein's ideas. As a result, a man who abhorred war and all kinds of violence, was castigated as an apostate and a traitor to the cause of peace.

Einstein's unorthodoxy was also evident in his attitude towards the Soviet Union. As a socialist he was initially sympathetic to Communist ideals and the October Revolution. But he changed his mind when the iniquities of Stalin's regime became apparent. He never visited Russia despite many invitations.

Thus it came about that the individual who personified the quest for global peace was criticized and vilified from all sides of the political spectrum. By the left for being pro-armament, by the right for propagating leftist views. This was particularly the case in the United States, where the FBI built up a huge dossier on Einstein as a possible Communist agitator or perhaps a spy. Even his scientific work came in for political criticism: in Nazi Germany his theories were banned as 'Jewish physics.'

Actually, his anti-war activities were inspired simply by his fervent—some might say, religious—reverence for the sanctity of life. He said:

> *My pacifism is an instinctive feeling, a feeling that possesses me; the thought of murdering another human being is abhorrent to me. My attitude is not the result of an intellectual theory, but is caused by a deep antipathy to every kind of cruelty and hatred...To me the killing of any human being is murder; it is also murder when it takes place on a large scale as an instrument of state policy.*

Apart from the loss of life, Einstein blamed war for the damage to cultural values in society: "War constitutes the most formidable obstacle to the growth of international cooperation, especially in its effect upon culture. War destroys all those conditions which are indispensable to the intellectual, if he is to work creatively."

Einstein's public anti-war activities began during the First World War. He was angered by the "Manifesto to the cultured world" issued in October 1914 by 93 German intellectuals. It attempted not only to whitewash the military atrocities by Germany, such as the violation of Belgium's neutrality, but to justify militarism as essential to German culture: "Were it not for German militarism, German culture would have been wiped off the face of the Earth." A strong riposte, under the title "Manifesto to the Europeans" eventually appeared. But only four intellectuals in Germany were willing to sign it, one of them being Einstein.

In the chauvinistic atmosphere of the time, a call for peace was seen as treason, whereas for Einstein it was a compelling challenge. He threw himself whole-

Einstein at his Long Island home with Leo Szilard in 1946, re-enacting for the television programme *March of Time* their famous meeting of 1939 that led to Einstein's letter to President Roosevelt.

heartedly into anti-war campaigns, co-founding some and lending his name to others. One of these, the Bund Neues Vaterland, also had a long-term objective, to establish a supranational organization which would make war impossible. Indeed, as the international situation became increasingly menacing in the 1930s, Einstein's thoughts were ever more focused on achieving a world government system. He did not visualize it as a replacement of existing national governments, but rather as a body with a specific aim: to prevent war by providing the means for solving disputes through negotiation. However, this objective required the relinquishing by national governments of some of their sovereignty—a step which was strongly advocated by Einstein.

Since the development of the atomic bomb during the Second World War, Einstein has often been referred to as its 'father,' because the basic principle of the bomb was the equivalence of mass and energy, as formulated by him in 1905. He himself did not foresee this application, but he did play a role in

Bertrand Russell, co-signatory of the Russell-Einstein manifesto, speaks at a "Ban the Bomb" demonstration in Trafalgar Square, London, 1961.

the bomb's development out of his fear that German scientists would produce the weapon and enable Hitler to win the war. In August 1939 he wrote a letter to President Roosevelt, acquainting him with the threat and urging that the United States start research on the military applications of nuclear fission; and in March 1940 he wrote again to Roosevelt. He was invited to become a member of an advisory committee on uranium, which he declined to join. This was the limit of his involvement in the atomic bomb project, which he later deeply regretted. "Had I known that the Germans would not succeed in developing the atomic bomb, I would not have supported its construction," he told the press in 1947. It is debatable whether Einstein's contribution was highly significant. The American effort did not start in earnest until 1942, with the setting up of the Manhattan Project. Apparently this step was triggered by information from Britain about work there that had established the bomb's scientific feasibility.

In 1945 the bomb was used to destroy Hiroshima and Nagasaki, despite the strong opposition of many scientists working on the Manhattan Project. Thereafter Einstein spent the remaining years of his life on efforts to prevent any further use of nuclear weapons. He was very active in movements such as the Emergency Committee of Atomic Scientists in the US, which advocated the international control of atomic energy and ultimately the abolition of nuclear arsenals.

The onset of the Cold War, and the nuclear arms race, with a greatly increased threat to humankind after the testing of the first hydrogen bomb in 1952, convinced Einstein of the urgent need for scientists to work together at stopping a nuclear holocaust. He laboured on this literally to the last day of his life, when he endorsed a proposal from the philosopher Bertrand Russell to bring together eminent scientists on both sides of the Iron Curtain to meet at a conference in order to seek solutions to the critical situation. Einstein's reply, which reached Russell after Einstein's death on 18 April 1955 was: "Thank you for your letter of April 5. I am gladly willing to sign your excellent statement. I also agree with your choice of the prospective signers. With kind regards, A. Einstein."

After the signatures of nine more scientists had been obtained, the statement, which became known as the "Russell-Einstein manifesto," was issued at a large press conference held at Caxton Hall in London on 9 July 1955. It generated publicity all over the world; and the direct outcome was the setting up of a new international movement of scientists which became known—from the name of the village in Canada where the first conference was held—as the Pugwash Conferences on Science and World Affairs.

The Pugwash movement played a very important role during the Cold War years, when discussions at its conferences facilitated the reaching of agreements such as the Nuclear Non-Proliferation Treaty, which helped to slow down the nuclear arms race and, later, to bring it to an end. For its contributions, Pugwash, jointly with its then-president, myself, the last survivor of the signatories of the Russell-Einstein manifesto, were awarded the Nobel peace prize in 1995.

Now, 50 years after Einstein's death, the threat of a nuclear holocaust is still with us. Indeed it has increased in the past few years. The warning in the final paragraph of the 1955 manifesto is as valid today as it was then:

> *There lies before us, if we choose, continual progress in happiness, knowledge and wisdom. Shall we, instead, choose death, because we cannot forget our quarrels? We appeal, as human beings, to human beings: Remember your humanity and forget the rest. If you can do so, the way lies open to a new paradise; if you cannot, there lies before you the risk of universal death.*

Leo Szilard addresses a meeting of the Pugwash movement.

14. The End of an Era

"Is there not a certain satisfaction in the fact that natural limits are set to the life of the individual, so that at its conclusion it may appear as a work of art?"

Einstein, 1947

Two weeks before his death, Einstein gave a delightfully lively and humorous interview to the historian of science I. Bernard Cohen, which is reprinted in this book. Since both of them were fascinated by Isaac Newton—his personality as well as his work—much of the talk was about him. At one point Einstein remarked in admiration of his seventeenth-century predecessor that "everything that Newton ever wrote is alive in the later works in physical science"—including, of course, in his own works.

A week or two before this interview, he had written to the family of his oldest friend Michele Besso, on hearing of his death at the age of 82: "Now he has departed from this strange world a little ahead of me. That signifies nothing. For us believing physicists, the distinction between past, present and future is only a stubbornly persistent illusion."

His own death, on 18 April 1955, was greeted by the world with a comparable certainty about Einstein's immortality. The front page of the *New York Times* carried tributes from the presidents of the United States and West Germany, and the prime ministers of Israel, France and India. Jawaharlal Nehru, the Indian prime minister, called Einstein "The great scientist of our age...truly a seeker after truth who would not compromise with evil or untruth." Moshe Sharett, the Israeli prime minister, said: "A powerful searchlight of the human mind, piercing by its rays the darkness of the unknown, has suddenly been extinguished. The world has lost its foremost genius and the Jewish people its most illustrious son in the present generation;" he was "the brightest jewel in their crown," said the widow of Chaim Weizmann, Israel's first president.

Intellectuals who had known Einstein personally chimed in with heartfelt tributes.

Robert Oppenheimer, the director of the Institute for Advanced Study: "He was one of the greats of all ages. For all scientists and most men, this is a day of mourning."

Bertrand Russell, who had just obtained Einstein's agreement to the Russell-Einstein manifesto: "Of all the public figures that I have known, Einstein was the one who commanded my most wholehearted admiration...

Einstein with Margot, Marguerite Wyler and his dog, Chico.

Einstein in his study, Princeton, 1954.

Einstein was not only a great scientist, he was a great man. He stood for peace in a world drifting towards war. He remained sane in a mad world, and liberal in a world of fanatics."

Niels Bohr, who had fundamentally disagreed with Einstein about quantum theory:

> *Through Albert Einstein's work the horizon of mankind has been immeasurably widened, at the same time as our world picture has attained a unity and harmony never dreamed of before. The background for such achievement was created by preceding generations of the worldwide community of scientists and its full consequences will only be revealed to coming generations.*
>
> *The gifts of Einstein are in no way confined to the sphere of science. Indeed his recognition of hitherto unheeded assumptions in even our most elementary and accustomed assumptions means to all people a new encouragement in tracing and combating the deep-rooted prejudices and complacencies inherent in every national culture.*

A few months later, in August 1955, a new chemical element was formally named einsteinium in his honour. Somewhat ironically, it had been produced as a result of the first test of the hydrogen bomb in 1952, which Einstein had strongly opposed. A second element was named fermium after the also recently deceased Enrico Fermi, a key player in the Manhattan Project. Both Einstein and Fermi, said the discoverers of the two new elements, had "played major roles in the birth of the atomic age."

As for Einstein's close friends, Maurice Solovine, who had kept in touch with him since the exhilarating days of the Olympia Academy in Bern half a century earlier, wrote:

> *I loved him and admired him profoundly for his basic goodness, his intellectual genius and his indomitable moral courage. In contrast to the lamentable vacillation that characterizes most so-called intellectuals, he fought tirelessly against injustice and evil. He will live in the memory of future generations not only as a scientific genius of exceptional stature but also as an epitome of moral greatness.*

For the general public, the cartoonist Herblock summed up the feeling in the *Washington Post* in a simple drawing of the heavenly spheres floating anonymously in space—except for one, the Earth, which carries a plaque with the words "ALBERT EINSTEIN LIVED HERE". Truly, Einstein was of cosmic significance.

In a local newspaper, the manager of the Institute for Advanced Study at Princeton recalled amusingly what this public adulation had meant for him as a manager. For many years he had tried to protect Einstein from the "hero worshippers and cranks who in their own peculiar way were every bit as resourceful as bobby soxers in full chase of some current idol of show business." As Einstein's copious and curious correspondence files prove, the public around the world regarded him, (the manager also wrote),

> *not only as the greatest of mathematicians, but as a statesman, philosopher, an oracle, and a symbol—and for good measure an authority on subjects as far removed as art, astrology and (on one occasion) old bones. They attempted to reach him by telephone at any hour of the day or night. They deluged him with mail. And they sought him out in person, arriving at the institute by bus, train, automobile and plane. Given half a chance they would sniff along the institute corridors like bloodhounds attempting to uncover his study which, as a matter of fact, was concealed behind two thin, unmarked, oak doors at one end of the ground floor.*

Almost the only voices not to join the chorus of praise were Einstein's family. He had outlived both his wives Mileva and Elsa, and his only sister Maja had died four years before. One of his children, Eduard, was a mental patient in Switzerland. The other, Hans Albert, was somewhat estranged from his father and had a poor relationship with Einstein's ever-faithful secretary Helen Dukas. Sad to record, more than 48 hours passed between the moment his father collapsed on 12 April and was taken to hospital until the moment Hans Albert was telephoned in California by an ill Margot Einstein and asked to come to Princeton. But when Hans Albert arrived, Einstein appeared pleased

Einstein walking on the streets of Princeton, early 1950s.

to see his son and chatted to him about science. However, neither at the time of Einstein's death nor in the years to follow did Hans Albert make any direct public statement about his father, though he did approve the text of a book (Peter Michelmore's *Einstein: Profile of the Man*, published in 1962) which was based on private conversations with Hans Albert. Most probably all children with a famous parent suffer from conflicting feelings about their relationship. In Hans Albert's case, the inner conflict seems to have been especially acute. Just before his death in 1973, he finally remarked of his father: "Probably the only project he ever gave up on was me. He tried to give me advice, but he soon discovered that I was too stubborn and that he was just wasting his time."

During Einstein's final interview, Cohen noted "no sense of the imminence of death." Einstein was extraordinarily alert and witty. But a week or so later, with the sudden collapse of his health, he knew he had not long to live—and he welcomed it. He had always been determined not to prolong his life through medical treatments. "I want to go when *I* want," he told his secretary. "It is tasteless to prolong life artificially. I have done my share; it is time to go. I will do it elegantly." He did not want his end to be "as in Haydn's *Farewell Symphony*, where one instrument of the orchestra vanishes after another"—as he had once characteristically tried to console a physicist friend who had lost his father unexpectedly. And he got his way with death, dying within less than a week of his collapse, still working away in his hospital bed at calculations for the unified theory; "completely in command of himself with regard to his condition," as Margot Einstein wrote to Hedwig Born soon after.

Einstein had asked to be cremated. He also wished for no religious rituals. The funeral took place immediately after his death, just six hours after the public announcement, and was attended by only his family and a very few others who had been close to him. Some lines written by Goethe for his friend Schiller's funeral service were read by one of Einstein's friends, Otto Nathan—nothing else. Then the ashes were taken and scattered nearby at an undisclosed spot. Einstein had made it very clear that he did not wish to leave anything of his physical body behind, to be "worshipped," as he wryly said, like the "bones of a saint."

Einstein (opposite) on his porch in Princeton shortly before his 75th birthday, 1954, and (below) in his study, 1953.

Einstein's Last Interview

I. Bernard Cohen

On a Sunday morning in April, two weeks before the death of Albert Einstein, I sat and talked with him about the history of scientific thought and great men in the physics of the past.

I had arrived at the Einstein home, a small frame house with green shutters, at 10 o'clock in the morning and was greeted by Helen Dukas, Einstein's secretary and housekeeper. She conducted me to a cheerful room on the second floor at the back of the house. This was Einstein's study. It was lined on two walls with books from floor to ceiling and contained a large low table laden with pads of paper, pencils, trinkets, books and a collection of well-worn pipes. There was a phonograph and records. Dominating the room was a large window with a pleasant green view. On the remaining wall were portraits of the two founders of the electromagnetic theory—Michael Faraday and James Clerk Maxwell.

After a few moments Einstein entered the room and Miss Dukas introduced us. He greeted me with a warm smile, went into the adjacent bedroom and returned with his pipe filled with tobacco. He wore an open shirt, a blue sweat shirt, grey flannel trousers and leather slippers. There was a touch of chill in the air, and he tucked a blanket around his feet. His face was contemplatively tragic and deeply lined, and yet his sparkling eyes made him seem ageless. His eyes

Einstein with the logician Kurt Gödel, Princeton, 1954.

watered almost continually; even in moments of laughter he would wipe away a tear with the back of his hand. He spoke softly and clearly; his command of English was remarkable, though marked by a German accent. The contrast between his soft speech and his ringing laughter was enormous. He enjoyed making jokes; every time he made a point that he liked, or heard something that appealed to him, he would burst into booming laughter that echoed from the walls.

Einstein in his study at Princeton, 1944.

We sat side by side at the table, facing the window and the view. He appreciated that it was difficult for me to begin a conversation with him; after a few moments he turned to me as if answering my unasked questions, and said: "There are so many unsolved problems in physics. There is so much that we do not know; our theories are far from adequate." Our talk veered at once to the problem of how often in the history of science great questions seem to be resolved, only to reappear in new form. Einstein expressed the view that perhaps this was a characteristic of physics, and suggested that some of the fundamental problems might always be with us.

Einstein remarked that when he was a young man the philosophy of science was considered a luxury, and most scientists paid no attention to it. He assumed that the situation was much the same with respect to the history of science. The two subjects must be similar, he said, because both deal with scientific thought. He wanted to know about my training in science and in history, and how I had become interested in Newton. I told him that one of the aspects of my research was the origin of scientific concepts and the relation between experiment and the creation of theory; what had always impressed me about Newton was his dual genius—in pure mathematics and mathematical physics and in experimental science. Einstein said that he had always admired Newton. As he explained this, I remembered those striking words in his autobiographical statement following a critique of Newtonian concepts—"Newton, forgive me."

Einstein was particularly interested in the various aspects of Newton's personality and we discussed Newton's controversy with Hooke in the matter of priority in the inverse-square law of gravitation. Hooke wanted only "some mention" in the preface to Newton's *Principia*, a little acknowledgment of his efforts, but Newton refused to make the gesture. Newton wrote to Halley, who was supervising the publication of the great *Principia*, that he would not give Hooke any credit; he would rather suppress the crowning glory of the treatise, the third and final "book" dealing with the system of the world. Einstein said:

"That, alas, is vanity. You find it in so many scientists. You know, it has always hurt me to think that Galilei did not acknowledge the work of Kepler."

We then spoke of Newton's controversy with Leibniz over the invention of the calculus, and how Newton had attempted to prove that his German contemporary was a plagiarist. There was set up a supposed international committee of inquiry, composed of Englishmen and two foreigners; today we know that Newton anonymously directed the committee's activities. Einstein said that he was shocked by such conduct. He did not appear too much impressed when I asserted that it was the nature of the age to have violent controversies, that the standards of scientific behaviour had changed greatly since Newton's day. Einstein felt that whatever the temper of the time there is a quality of human dignity that should enable a man to rise above the passions of his age.

Then we talked about Franklin, whose conduct as a scientist I had always admired, especially because he had not entered into such controversy. Franklin was proud that he had never written a polemic in defence of his experiments or his ideas. He believed that experiments can be tested only in the laboratory, and that concepts and theories must make their own way by proving their validity. Einstein only partly agreed. It was well to avoid personal fights, he said, but it was also important for a man to stand up for his own ideas. He should not simply let them go by default, as if he did not really believe in them.

Einstein, who knew of my interest in Franklin, wanted to know more about him: Had he done more in science than invent the lightning rod? Had he really done anything of importance? I replied that in my opinion the greatest thing to come out of Franklin's research was the principle of the conservation of charge. Yes, said Einstein, that was a great contribution. Then he thought to himself for a moment or two and, with a smile, asked me how Franklin could have proved it. Of course, I conceded, Franklin was only able to adduce some experimental examples of equal positive and negative electrification, and to show the applicability of the principle in explaining a variety of phenomena. Einstein shook his head once or twice, and admitted that until then he had not appreciated that Franklin deserved a place of honour in the history of physics.

The subject of controversies over scientific work led Einstein to take up the subject of unorthodox ideas. He mentioned a fairly recent and controversial book, of which he had found the non-scientific part—dealing with comparative mythology and folklore—interesting. "You know," he said to me, "it is not a bad book. No, it really isn't a bad book. The only trouble with it is, it is crazy." This was followed by a loud burst of laughter. He then went on to explain what he meant by this distinction. The author had thought he was basing some of his ideas upon modern science, but found the scientists did not agree with him at all. In order to defend his idea of what he conceived modern science to be, so as to maintain his theories, he had to turn around and attack the scientists. I replied that the historian often encountered this problem: can a scientist's contemporaries tell whether he is a crank or a genius when the only evident fact is his unorthodoxy? A radical like Kepler, for example, challenged accepted ideas; it must have been difficult for his contemporaries to tell whether he was a genius or a crank. "There is no objective test," replied Einstein.

Einstein was sorry that scientists in the United States had protested to publishers about the publication of such a book. He thought that bringing pressure to bear on a publisher to suppress a book was an evil thing to do. Such a book really could not do any harm, and was therefore not really bad. Left to itself, it would have its moment, public interest would die away and that would be the end of it. The author of such a book might be "crazy" but not "bad," just as the book was not "bad." Einstein expressed himself on this point with great passion.

Much of the time we spent together was devoted to the history of science, a subject that had long been of interest to Einstein. He had written many articles about Newton, prefaces to historical works and also biographical sketches of his contemporaries and the great men of science of the past. Thinking aloud about the nature of the historian's job, he compared history to science. Certainly, he said, history is less objective than science. For example, he explained, if two men were to study the same subject in history, each would stress the particular part of the subject which interested him or appealed to him the most. As Einstein saw it, there is an inner or intuitional history and an external or documentary history. The latter is more objective, but the former is more interesting. The use of intuition is dangerous but necessary in all kinds of historical work, especially when the attempt is made to reconstruct the thought processes of someone who is no longer alive. This kind of history, Einstein felt, is very illuminating despite its riskiness.

It is important to know, he went on, what Newton thought and why he did certain things. We agreed that the challenge of such a problem should be the major motivation of a good scientific historian. For instance, how and why had

Newton developed his concept of the ether? Despite the success of Newton's gravitation theory, he was not satisfied by the concept of the gravitational force. Einstein believed that what Newton most strongly objected to was the idea of a force being able to transmit itself through empty space. Newton hoped by means of an ether to reduce action at a distance to a force of contact. Here is a statement of great interest about Newton's process of thought, Einstein declared, but the question arises as to whether—or perhaps to what extent—one can document such intuition. Einstein said most emphatically that he thought the worst person to document any ideas about how discoveries are made is the discoverer. Many people, he went on, had asked him how he had come to think of this or how he had come to think of that. He had always found himself a very poor source of information concerning the genesis of his own ideas. Einstein believed that the historian is likely to have a better insight into the thought processes of a scientist than the scientist himself.

Einstein's interest in Newton had always been centred on his ideas, which are to be found in every textbook of physics. He had never made a systematic examination of all Newton's writings, in the manner of a thorough historian of science, but of course he had an appreciation of Newtonian science that could come only from a scientific peer of Newton. Yet Einstein was keenly interested in the results of scholarship in the history of science, such as the development of some of Newton's fundamental opinions in his successive revisions of his major works, the *Opticks* and the *Principia*. In our correspondence on this subject, the question had arisen as to whether there was any sense in which Einstein might have 'revived' a Newtonian concept of light in his paper on photons in 1905. Had he ever read Newton's writings on light before that year? He told me: "As far as I can remember I had not studied, or at least not studied profoundly, the original before I had to write the little foreword for the *Opticks*. The reason is, of course, that everything that Newton ever wrote is alive in the later works in physical science." Furthermore, "younger people are very little historically minded." Einstein's main concern had been his own scientific work; he had known of Newton primarily as the author of many of the fundamental concepts in classical physics. But he had encountered Newton's "utterances of a philosophical character"; these were cited again and again.

In 1905 Einstein knew that Newton had espoused a corpuscular theory of light, a fact which he must have found in Drude's famous book on light, but he had evidently not known until many decades later about Newton's attempts to blend a corpuscular and wave theory. Einstein knew of my interest in the *Opticks*, especially in the influence of this book on the later course of experimental physics. When I remarked on the greatness of Newton's intuition about the study of light being the key to exact knowledge of the corpuscles of matter, Einstein misunderstood what I had said. He replied that we must not take too seriously the historical accident that Newton's corpuscular view of light with wave aspects sounds something like modern statements. I explained what I had meant: Newton had attempted to infer from what we call interference or diffraction phenomena the size of the corpuscles of matter. These intuitions might be very profound, Einstein agreed, but not necessarily fruitful. For example, he said, Newton's thoughts on this subject did not lead anywhere; he could not prove his point nor derive precise information about the structure of matter.

Einstein in his office in Princeton, early 1950s, and (opposite) the same office, 1970.

Einstein was actually more interested in the *Principia* and in Newton's views on hypotheses. He greatly esteemed the *Opticks*, but primarily for the analysis of colour and the magnificent experiments. Of this book he had written that "it

alone can afford us the enjoyment of a look at the personal activity of this unique man." Looking back over all of Newton's ideas, Einstein said, he thought that Newton's greatest achievement was his recognition of the role of privileged systems. He repeated this statement several times and with great emphasis. This is rather puzzling, I thought to myself, because today we believe that there are no privileged systems, only inertial systems; there is no privileged frame—not even our solar system—which we can say is privileged in the sense of being fixed in space, or having special physical properties not possible in other systems. Due to Einstein's own work we no longer believe (as Newton did) in concepts of absolute space and absolute time, nor in a privileged system at rest or in motion with respect to absolute space. Newton's solution appeared to Einstein ingenious and necessary in his day. I was reminded of Einstein's statement: "Newton,...you found the only way which, in your age, was just about possible for a man of highest thought and creative power."

I remarked that Newton's genius was displayed in his adopting as a "hypothesis" in the *Principia* the statement about the "centre of the system of the world" being

Einstein in Princeton, 1953.

fixed, immobile in space; that a lesser man than Newton might have thought he could prove such an assertion, either by mathematics or by experiment. Einstein replied that Newton probably did not fool himself. He was apt to know what could be proved and what could not; this was a sign of his genius.

Einstein then said that the biographical aspects of scientists had always interested him as much as their ideas. He liked to learn the lives of the men who had created the great theories and performed the major experiments, what kind of men they were, how they worked and how they treated their fellow men. Reverting to an earlier topic of our conversation, Einstein observed how many scientists seemed to have suffered from vanity. He pointed out that vanity may appear in many different forms. Often a man would say that he had no vanity, but this too was a kind of vanity because he took such special pride in the fact. "It is like childishness," he said. Then he turned to me and his booming laugh filled the room. "Many of us are childish; some of us more childish than others. But if a man knows that he is childish, then that knowledge can be a mitigating factor."

The conversation then turned to Newton's life and his private speculations: his investigations of theology. I mentioned to Einstein that Newton had essayed a linguistic analysis of theology, in an attempt to find the corruptions that had been introduced into Christianity. Newton was not an orthodox Trinitarian. He believed his own views were hidden away in Scripture, but that the revealed documents had been corrupted by later writers who had introduced new concepts and even new words. So Newton sought by linguistic analysis to find the truth. Einstein remarked that for him this was a "weakness" in Newton. He did not see why Newton, finding his own ideas and the orthodox ones at variance, did not simply reject the established views and assert his own. For instance, if Newton could not agree with the accepted interpretations of Scripture, why did he believe that Scripture must nevertheless be true? Was it only because the common point of view was that fundamental truths are contained in the Bible? It did not seem to Einstein that in theology Newton showed the same great quality of mind as in physics. Einstein apparently had little feeling for the way in which a man's mind is imprisoned by his culture and the character of his thoughts are moulded by his intellectual environment. I did not press the point, but I was struck by the fact that in physics Einstein could see Newton as a man of the seventeenth century, but that in the other realms of thought and action he viewed each man as a timeless, freely acting individual to be judged as if he were a contemporary of ours.

Einstein seemed particularly impressed by the fact that Newton had not been entirely satisfied with his theological writings, and had sealed them all up in a box. This seemed to indicate to Einstein that Newton was aware of the imperfect quality of his theological conclusions and would not present to public view any writings that did not measure up to his own high standards. Since Newton obviously did not wish to publish his speculations on theology, Einstein asserted with some passion that he personally hoped no one else would publish them. Einstein said a man has a right to privacy, even after his death. He praised the Royal Society for having resisted all pressure to edit and print writings of Newton which their author had not wanted to publish. He believed that Newton's correspondence could justly be published, because a letter written and sent was intended to be read, but he added that even in correspondence there might be some personal things which should not be published.

Then he spoke briefly about two great physicists whom he had known well: Max Planck and H. A. Lorentz. Einstein told me how he had come to know Lorentz in Leiden through Paul Ehrenfest. He remarked that he had admired and loved Lorentz perhaps more than anyone else he had ever known, and not only as a scientist. Lorentz had been active in the movement for "international cooperation," and had always been interested in the welfare of his fellow men. He had worked on many technical problems for his own country, an activity which was not generally known. This was part of Lorentz's character, Einstein explained, a kind of nobility which made him work for the well-being of others, preferably in anonymity. Einstein also expressed great affection for Max Planck. Planck was a religious man, he said, and always sought to reintroduce the absolutes—even on the basis of relativity theory. I asked Einstein whether Planck had ever fully accepted the "theory of photons," or whether he had continued to restrict his interest to the absorption or emission of light without regard to its transmission. Einstein stared at me for a moment or two in silence. Then he smiled and said: "No, not a theory. Not a *theory* of photons," and again his deep laughter enveloped us both—and the question was never answered. I remembered that Einstein's 1905 paper, for which (nominally) he had been awarded the Nobel prize, did not contain the word 'theory' in the title, but instead referred to considerations from a "heuristic viewpoint."

There are fashions in science, Einstein said. When he had studied physics as a young man, one of the major questions being discussed was: do molecules exist? He remembered how important scientists, men like Wilhelm Ostwald and Ernst Mach, had been explicit in stating that they did not really believe in atoms and molecules. One of the greatest differences between physics then and now, Einstein observed, was that today nobody bothers to ask this particular question any more. Although Einstein did not agree with the radical position adopted by Mach, he told me he admired Mach's writings, which had had a great influence on him. He had visited Mach, he said, in 1913, and had raised a question in order to test him. He asked Mach what his position would be if it proved possible to predict a property of a gas by assuming the existence of atoms—some property that could not be predicted without the assumption of atoms and yet one that could be observed. Einstein said he had always believed that the invention of scientific concepts and the building of theories upon them was one of the great creative properties of the human mind. His own view was thus opposed to Mach's, because Mach assumed that the laws of science were only an economical way of describing a large collection of facts. Could Mach accept the hypothesis of atoms under the circumstances Einstein had stated, even if it meant very complicated computations? Einstein told me how delighted he was when Mach replied affirmatively. If an atomic hypothesis would make it possible to connect by logic some observable properties which would remain unconnected without this hypothesis, then, Mach said, he would have to accept it. Under these circumstances it would be "economical" to

Einstein at his home in Princeton. "Dominating the room was a large window with a pleasant green view," wrote I. Bernard Cohen.

assume that atoms may exist because then one could derive relations between observations. Einstein had been satisfied; indeed more than a little pleased. With a serious expression on his face, he told me the story all over again to be sure that I understood it fully. Wholly apart from the philosophical victory over what Einstein had conceived Mach's philosophy to have been, he had been gratified because Mach admitted that there might, after all, be some use to the atomistic philosophy to which Einstein had been so strongly committed.

Einstein said that at the beginning of the century only a few scientists had been philosophically minded, but today physicists are almost all philosophers, although "they are apt to be bad philosophers." He pointed as an example to logical positivism, which he felt was a kind of philosophy that came out of physics.

Now it was time to leave. I was horrified to realize it was a quarter to twelve. Knowing that Einstein tired easily, I had meant to stay only half an hour. Yet every time I had got up to depart he had said, "No, no, don't go yet. You have come to see me about your work and there is still more to talk about." Yet at last I was taking my leave. Miss Dukas joined us as we walked toward the front of the house. As I neared the stairs, I turned to thank Einstein, missed a step and almost fell. When I had recovered my balance, Einstein smiled and said, "You must be careful here, the geometry is complicated. You see," he continued, "negotiating stairs is not really a physical problem, but a problem in applied geometry." He chuckled and then laughed out loud. I started down the stairs

Einstein with Chico and Helen Dukas.

and Einstein began to walk down the corridor toward the study. Suddenly he turned and called: "Wait. Wait. I must show you my birthday present."

As I returned to the study Miss Dukas explained to me that Eric Rogers, who teaches physics at Princeton, had made a gadget for Einstein as a present for his 76th birthday, and that Professor Einstein had been delighted with it. Back in the study, I saw Einstein take from the corner of the room what looked like a curtain rod five feet tall, at the top of which was a plastic sphere about four inches in diameter. Coming up from the rod into the sphere was a small plastic tube about two inches long, terminating in the centre of the sphere. Out of this tube there came a string with a little ball at the end. "You see," said Einstein, "this is designed as a model to illustrate the equivalence principle. The little ball is attached to a string, which goes into the little tube in the centre and is attached to a spring. The spring pulls on the ball, but it cannot pull the ball up and into the little tube because the spring is not strong enough to overcome the gravitational force which pulls down on the ball." A big grin spread across his face and his eyes twinkled with delight as he said: "And now the equivalence principle." Grasping the gadget in the middle of the long brass curtain rod, he thrust it upward until the sphere touched the ceiling. "Now I will let it drop," he said, "and according to the equivalence principle there will be no gravitational force. So the spring will now be strong enough to bring the little ball into the plastic tube." With that he suddenly let the gadget fall freely and vertically, guiding it with his hand, until the bottom reached the floor. The plastic sphere at the top was now at eye level. Sure enough, the ball nestled in the tube.

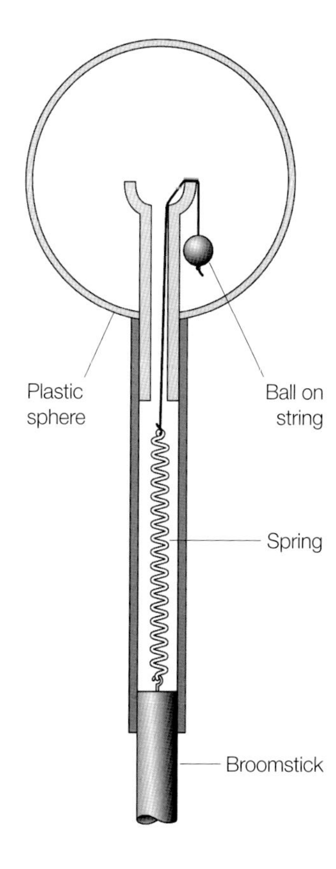

Fig. 8: A representation of the gadget made for Einstein by his colleague Eric Rogers to demonstrate the equivalence principle, and which Einstein showed Cohen during his last interview. Rogers indicates that a "broomstick" was used to make the gadget—rather than the curtain rod imagined by Cohen.

With the demonstration of the birthday present our meeting was at an end. As I walked out to the street, I thought to myself that of course I had known that Einstein was a great man and a great scientist, but I had had no idea of the warmth of his friendly personality, his kindness and his rich sense of humour.

There had been, during that visit, no sense of the imminence of death. Einstein's mind was alert, his wit was keen and he had seemed very gay. On the Saturday following my visit, a week before Einstein was taken to the hospital, a Princeton friend of long standing and intimacy went with Einstein to the hospital to see Einstein's daughter, who was ill with sciatica. This friend writes that after he and Einstein left the hospital that Saturday, "we went for a long walk. Strange to say, we talked about our attitudes toward death. I mentioned a quotation from James Frazer in which he said that fear of death was the basis of primitive religion, and that to me death was both a fact and a mystery. Einstein added, 'And also a relief.'"

15. Einstein's Enduring Magic

"Knowledge exists in two forms—lifeless, stored in books, and alive in the consciousness of men. The second form of existence is after all the essential one; the first, indispensable as it may be, occupies only an inferior position."

Einstein, "Message in honour of Morris Raphael Cohen," 1949

Einstein never wanted to be an icon or a saint. Hence his decision to be cremated and have his ashes scattered. He also asked that his house at 112 Mercer Street in Princeton should never be turned into a museum and become what he called "a place of pilgrimage." Just how right he was to be concerned about his deification is demonstrated by the posthumous saga of his brain.

He unambiguously refused a request that his brain should be removed and examined after his death. But in the event Thomas Harvey, the pathologist who did the post-mortem, ignored Einstein's wishes. Though he was not a trained neuroscientist, Harvey became obsessed with the brain and kept hold of bits of it for decades before finally returning them to a laboratory in Princeton in his late eighties. He thought that the brain would "reveal the secrets of genius, and that it would make him famous," writes a current Cambridge University professor of neuroscience, Joe Herbert. "Neither happened. But the brain carried with it the charisma of Einstein himself, so that everyone who saw it was galvanized by thoughts of wealth or glory. Scientists, journalists, entrepreneurs, Einstein's executors—all tried to grab a piece of the action." Willy-nilly the organ had become like a saint's relic: instead of a hair, some blood or a toenail, there was a mass of pickled brain tissue. Yet the fact is, since 1955 the only mildly interesting piece of scientific research to have emerged from studying Einstein's brain is that a part of it—the part that, among other things, relates to a person's mathematical ability—is bigger than the same part in a collection of other brains.

The scientific fascination with Einstein's brain, if unproductive, is understandable. Genius is quite as compelling a subject for scientists as it is for everyone else, if not more so. The biologist Richard Dawkins—who is not known for his intellectual humility—described himself recently as "unworthy to lace Einstein's sockless shoes." How do we recognize genius and where does it come from—and how can we increase our own thinking power and that of our offspring? Not by formal education, judging from the experience of Einstein, who as we know was induced to leave his German high school for being a disruptive teenager. Ah yes, say parents—but then *he* was a genius.

Einstein has become a byword for brainpower. A full-page advertisement in *Nature*, one of the world's two leading science journals, consisted solely of a well-known quotation supposedly from Einstein: "Things should be made as simple as possible *but not any simpler*." The idea was to promote *Nature*'s news coverage of the latest science as being not specialized, but not sensationalist either. The *New York Times* chose an

Einstein on his 74th birthday at Princeton Inn, 1953.

Einstein with a bust of himself made by the sculptor Gina Plunguian. She was also known to step in and help out as Einstein's secretary when Helen Dukas was away.

image of Einstein to promote its online search engine. A recent book by a computer graphics expert is entitled *Thinking Like Einstein*—the key idea being to improve computer graphics with more reliance on image-based thinking *à la* Einstein and less on conventional word-based thinking. Elsewhere a publisher's advertisement showed three identical Einsteins side by side and asked, "…If cloning could produce several Einsteins, would you approve its use?" And, for a laugh, a cartoon strip in *Scientific American* about *your* chance of becoming a "Sudden Genius" showed a sketch of a book called *The Einstein Diet* with the caption: "What did this mega-genius eat? Read this book and unlock Albert's diet secrets." A snip at $84.99.

It has to be said that the man himself, in his lifetime, gave full currency to today's abiding impression of him as an eccentric, lovable genius. There is an irresistible anecdote about Einstein in the act of thinking told by one of his assistants, Banesh Hoffmann, who worked with Einstein and Leopold Infeld in 1937–38 on general relativity and later became a professor of physics at the City University of New York. So irresistible, that it is worth quoting in full:

> *Whenever we came to an impasse the three of us had heated discussions—in English for my benefit, because my German was not too fluent—but when the argument became really intricate Einstein, without realizing it, would lapse into German. He thought more readily in his native tongue. Infeld would join him in that tongue, while I struggled so hard to follow what was being said that I rarely had time to interject a remark till the excitement died down.*
>
> *When it became clear, as it often did, that even resorting to German did not solve the problem, we would all pause, and then Einstein would stand up quietly and say, in his quaint English, 'I vill a little*

t'ink'. So saying he would pace up and down or walk around in circles, all the time twirling a lock of his long, greying hair around his forefinger. At these moments of high drama Infeld and I would remain completely still, not daring to move or make a sound, lest we interrupt his train of thought. A minute would pass in this way and another, and Infeld and I would eye each other silently while Einstein continued pacing and all the time twirling his hair. There was a dreamy, far-away, and yet sort of inward look on his face. There was no appearance at all of intense concentration. Another minute would pass and another, and then all of a sudden Einstein would visibly relax and a smile would light up his face. No longer did he pace and twirl his hair. He seemed to come back to his surroundings and to notice us once more, and then he would tell us the solution to the problem and almost always the solution worked.

So here we were, with the magic performed triumphantly and the solution sometimes was so simple we could have kicked ourselves for not having been able to think of it by ourselves. But that magic was performed invisibly in the recesses of Einstein's mind, by a process that we could not fathom. From this point of view the whole thing was completely frustrating. But, from the more immediately practical point of view, it was just the opposite, since it opened a way to further progress and without it we should never have been able to bring the research to a successful conclusion.

It is easy to understand Einstein's profound appeal to scientists, especially physicists, given the reach of his ideas as described in the first seven chapters of this book—though they should bear in mind Philip Anderson's warning that Einstein's achievement was by no means all the result of pure thought. Anyone interested in science is bound to be interested in Einstein. *Scientific American*, in its recent Einstein special issue, estimated that two thirds of the "crackpot missives" sent to scientists and science magazines relate to Einstein's theories. Either the writer claims to have found a unified theory where Einstein failed, or the claim is to have proved Einstein's ideas, especially relativity, wrong. (The other third of the missives concerns perpetual-motion machines and infinite-energy sources.)

But there must be much more to Einstein's appeal, which goes far beyond the world of science and scientists, than his great thinking power. Arthur C. Clarke—whose own writings and personality have reached beyond merely the readers and cinema goers who like science fiction—puts Einstein's enduring fame across the globe down to "the unique combination of genius, humanist, pacifist and eccentric," in the final article of this book. But Clarke admits that this is only his personal guess, not a full explanation of the Einstein phenomenon.

After all, Newton is a household name—perhaps the only scientist who is, apart from Einstein and Darwin. Yet how many advertisers would think of using his image to promote a product for the general public, except perhaps apples. No politician is likely to drop Newton's name in a speech, other than a speech about science. Newton is seldom quoted outside a scientific context. Of course Newton biographies continue to be written, but Newton does not pop up in newspaper headlines, cartoons and ordinary conversation much. There are only a handful of well-known anecdotes about Newton, and no Newton jokes. One cannot imagine a popular book entitled "The Quotable Newton"—unlike *The Quotable Einstein*, now in its third edition and still expanding. Some of its quotations, as in *Nature*'s one above, are of uncertain provenance but seem to have the authentic Einstein ring—such was his wit, variety and inimitability.

The reason for the difference in public estimation is that Newton is celebrated for his scientific achievements, for which all subsequent physicists, especially Einstein, continue to revere him; and so is Einstein, even if relativity and quantum theory are not as accessible to non-scientists as mechanics and optics. Unlike Newton, however, Einstein is appreciated as a unique and decent human being too.

Newton's loveless, orphaned childhood and his extreme unsociability as an adult, along with his acerbic controversies with scientific contemporaries such as Hooke and Leibniz and his scurvy treatment of the astronomer royal John Flamsteed, have become fairly widely known in the half-century since Einstein spoke of Newton in his interview with I. B. Cohen. However it is worth emphasizing one astonishing detail to underline how different Newton was from Einstein. After Newton departed Cambridge and moved to London in 1696 he left behind not a single friend in the place where he had spent 35 years and done his revolutionary work; there is not one surviving letter written by him to any of his Cambridge acquaintances between 1696 and his death in 1727. His successor as Lucasian professor of mathematics, William Whiston, wrote of Newton in his memoirs (long after his patron's death): "He was of the most fearful, cautious and suspicious temper, that I ever knew." As Jacob Bronowski rightly said in his book *The Ascent of Man*: "Newton is the Old Testament god; it is Einstein who is the New Testament figure...full of humanity, pity, a sense of enormous sympathy."

Einstein and Newton shared a great deal in their scientific work, but had very little in common as human beings. For all Einstein's scepticism about personal relationships and institutional life, his two unsuccessful marriages and his family tragedies, he was frequently highly sociable, a regular public speaker and kept up a vast correspondence with friends, colleagues and strangers. He made constant efforts to help scientific 'rivals' and newcomers, for example Satyendranath Bose, a name from India which was entirely unknown to him. And his disagreements over science and all other matters—except anti-Semitism and Nazism—were always conducted without polemic and usually without rancour. There is no malice even in his long and inconclusive battle with Niels Bohr over quantum theory. Einstein hit hard but not in order to wound. Arguing with his close friend Max Born on the same subject in the 1940s and 50s, Einstein was extremely firm in his opposition, but about the closest he came to a personal attack was the slightly sardonic comment, "Blush, Born, Blush!"

In addition to his personal qualities, almost all of the public causes Einstein supported were admirable and far-sighted. Many required courage in the conditions of his time. He stood up to be counted—and attacked—against discreet and rabid anti-Semitism, against the segregation and lynching of American blacks, against the witch-hunts of McCarthyism, against the build-up of the military-industrial complex and against the drift towards nuclear war, when none of these was a fashionable or 'respectable' issue. Instead

Einstein on Long Island, New York, where he kept a summer house.

of basking in his fame and enjoying himself with physics, music and sailing, he fought oppression wherever he thought his name and reputation might have a desirable impact. One cannot say that Einstein's intervention was decisive in defeating any of these tendencies, but there is ample testimony that he gave hope to the persecuted and influenced public debate. The very fact that J. Edgar Hoover was determined to 'get' Einstein shows how seriously Einstein's activism was taken by reactionary forces. Hoover's abject failure cannot but suggest that justice does eventually prevail over injustice (besides providing much unintentional entertainment at the bumbling ineptitude of the FBI in the 1950s).

No wonder Einstein was inspired by Mahatma Gandhi, although he did not accept Gandhi's view that civil disobedience should be tried against the Nazis. To a great extent he shared Gandhi's indifference to material success. In 1952, he called Gandhi "the greatest political genius of our time... He gave proof of what sacrifice man is capable once he has discovered the right path. His work on behalf of India's liberation is living testimony to the fact that man's will, sustained by an indomitable conviction, is more powerful than material forces that seem insurmountable."

Einstein's attitude to religion has been influential too. No one has yet launched a movement on the basis of Einstein's 'cosmic religion.' But his ideas are taken seriously across the religious spectrum and provoke debate. Richard Dawkins, well known for his atheism, thinks that "Einstein was profoundly spiritual, but he disowned supernaturalism and denied all personal gods... I gladly share his magnificently godless spirituality. No theist should presume to give Einstein lessons in spirituality." Stephen Hawking, though he does not refer explicitly to Einstein's religion, showed a similar outlook to Einstein's when he wrote in 1984 that, "It would be perfectly consistent with all we know to say that there was a Being who was responsible for the laws of physics. However, I think it could be misleading to call such a Being 'God,' because this term is normally understood to have personal connotations which are not present in the laws of physics." (In 1988, though, in his book *A Brief History of Time*, Hawking

suggested that human reason might ultimately come to know "the mind of God.") And Pope John Paul II, speaking on Einstein's birth centenary in 1979 to a meeting of the Pontifical Academy of Sciences, said:

> *Filled with admiration for the genius of the great scientist, in whom is revealed the imprint of the creative spirit, without intervening in any way with a judgement on the doctrines concerning the great systems of the universe, which is not in her power to make, the Church nevertheless recommends these doctrines for consideration by theologians in order to discover the harmony that exists between scientific truth and revealed truth.*

How influential has Einstein been on the arts? During his lifetime, Max Brod—the editor of Franz Kafka—wrote a novel, his most famous work, *Tycho Brahe's Way to God* (1915), in which the character of Kepler was closely based on Einstein, whom Brod came to know in Prague in 1911–12. Brod commented that Einstein "time and again filled me with amazement, and indeed enthusiasm, as I watched the ease with which he would, in discussion, experimentally

Einstein and Elsa arriving in America, 1930.

change his point of view, at times tentatively adopting the opposite view and viewing the whole problem from a new and totally changed angle." Other artists, such as the poets William Carlos Williams and E. E. Cummings, and the novelist Karel Capek, have mentioned Einstein in their works. In Bertolt Brecht's play *Galileo* (1943), Brecht called himself the "Einstein of the new theatrical form." Since Einstein's death there have been notable presentations of him in Friedrich Dürrenmatt's play *The Physicists* (1962), in Philip Glass's opera *Einstein on the Beach* (1976)—which Glass discussed earlier in this book—and in the physicist Alan Lightman's novel *Einstein's Dreams* (1993).

As for the more subtle influences of Einstein's ideas on artists, attempts have been made to link him with the works of, among other modernist writers who use multiple viewpoints, T. S. Eliot, Virginia Woolf and Lawrence Durrell. But as the authors of a study, *Einstein as Myth and Muse*, admit, there is no clinching evidence. Referring to Durrell's *Alexandria Quartet* (1957–60), Alan Friedman and Carol Donley comment honestly: "Simply because writers say they are using relativity...does not mean either that they understand it or that their adaptations of relativity principles succeed artistically." A whole book, *Einstein, Picasso: Space, Time, and the Beauty That Causes Havoc*, has been written by a historian of science, Arthur Miller, on the connections between Einstein's science and the paintings of Pablo Picasso. It makes the interesting point that although both Einstein and Picasso broke with classicism and adopted multiple viewpoints (in relativity theory and in cubism), neither of them could ever come to terms with a total abandonment of classicism; Einstein disliked the probabilistic basis of quantum theory and Picasso disliked totally abstract art. But since Einstein showed no recorded interest in Picasso's art, and Picasso no recorded interest in Einstein's science, the whole enterprise of comparing the two seems doomed to failure.

It is tempting to recall here Einstein's ironic comment on the philosophers of relativity: "the less they know about physics the more they philosophize." And perhaps also Paul Dirac's warning about trying to link science and art: "In science one tries to tell people, in such a way as to be understood by everyone, something that no one ever knew before. But in poetry, it's the exact opposite."

Einstein's own tastes in the arts seem to have been largely classical. In music, the art form he was most drawn to, his favourites were undoubtedly Bach and Mozart. It was Mozart's sonatas that first persuaded him, at the age of thirteen, to make the effort to learn how to play the violin, after he had rejected—ever the rebel against authority—the mechanical practice imposed on him as a child by a music teacher. He once said: "Mozart's music is so pure and beautiful that I see it as a reflection of the inner beauty of the universe." Einstein's name and science will surely be remembered as long as Mozart's name and music. He too searched for and reflected in his works the inner beauty of the universe.

DECEMBER 31, 1999 $4.95 www.time.com

PERSON OF THE CENTURY

TIME

ALBERT EINSTEIN

Einstein: Twentieth-Century Icon

Arthur C. Clarke

I never met Einstein in person, even though his theories interested me a great deal—both when I was studying physics and mathematics at King's College, London, in the late 1940s, and in my work with the British Interplanetary Society.

The closest our orbits came to crossing was in 1996, when the International Astronomical Union (IAU) wanted to name an asteroid in my honour. Because of my association with a certain movie, I enquired whether asteroid 2001 was still available. The IAU came back saying: sorry, it has already been assigned to one A. Einstein. But I was happy to become the absentee landlord of Asteroid Clarke, formerly No. 4923.

It is hard to imagine the world as we know it today without the influence of Einstein, who left his mark not only in the realm of big ideas such as the nature of time, the fate of the universe and the speed of light. From hand-held GPS units to digital cameras, and from lasers to solar-powered devices, a whole range of everyday gadgets has benefited from the intellectual legacy of the former patent clerk. (His Nobel prize in physics, for 1921, was mainly for his work on the photoelectric effect that led to the tapping of solar power.)

Half a century after his death, Einstein's iconic status remains as great as ever. It transcends cultural and geographical barriers—he is just as easily recognized in rural India or in remote Siberia as in Europe or America. Inevitably, he has also become the stereotype of the eccentric genius in countless movies and works of fiction.

Few people in history deserve the superstar status as much as Einstein does. In the space of a few years, he changed our perception and understanding of the physical world—and indeed the universe—just as Isaac Newton had done three centuries earlier. That previous transformation inspired Alexander Pope to propose an epitaph for Newton:

Nature, and Nature's laws lay hid in night:
God said, Let Newton be! and all was light.

By the early twentieth century, however, Newton's laws did not explain everything and it was Einstein who restored some order to the increasing confusion. But his theories were much more complicated, which prompted Sir John Squire to cap Pope's lines with:

It did not last: the Devil howling 'Ho!
Let Einstein be!' restored the status quo.

What social and cultural factors made Einstein the first global superstar of science? I will leave that to be analysed in some future PhD thesis. My own view is that the unique combination of genius, humanist, pacifist and eccentric made him accessible—and even loveable—to tens of millions of people, many of whom never understood his profound scientific theories. He was not only the most famous scientist of his time, but he made good use of his fame to support a variety of good causes.

During the last few years of his life, he intensified his personal campaign against thermonuclear weapons. I can still remember the global reverberations when the Russell-Einstein manifesto was published in London in the summer of 1955, only a few weeks after Einstein's death. It took extraordinary courage, even for the leading scientists of the time, to take this uncompromising stand against the new-found weapons of mass destruction. Einstein knew what he was talking about. He had cautioned in 1949: "I do not know how the Third World War will be fought, but I can tell you what they will use in the Fourth—rocks!"

These sentiments resonated strongly with me. Only a few years before this, as a young radar officer of the Royal Air Force, I had shared the global shock and horror of the first atomic bomb attacks that ended the war. In the weeks following the attacks on Hiroshima and Nagasaki, I had written one of my most important essays, "The rocket and the future of warfare," which ended with these words: "Upon us, the heirs to all the past and trustees of a future which our folly can slay before its birth, lies a responsibility no other age has ever known. If we fail in our generation those who come after us may be too few to rebuild the world when the dust of the cities has descended and the radiation of the rocks has died away."

Freeman Dyson began this book with one of Einstein's many paradoxes—about black holes. I would like to end it with one of my own, which I first articulated in the 1960s. It concerns God—a concept of interest to Einstein and to myself, for much the same reasons.

A few decades ago, I began to be worried by the following astro-theological paradox. It is hard to believe that no one else has ever thought of it, yet I have never seen it discussed anywhere.

One of the most firmly established facts of modern physics and the basis of Einstein's theory of relativity is that the velocity of light is the speed limit of the material universe. No object, no signal, no *influence*, can travel any faster than this. Please don't ask why this should be; the universe just happens to be built that way. Or so it seems at the moment.

But light takes not millions, but *billions* of years to cross even the part of creation we can observe with our telescopes. So, if God obeys the laws She apparently established, at any given time She can have control over only an infinitesimal fraction of the universe. All hell (literally?) might be breaking loose ten light years away, which is a mere stone's throw in interstellar space, and the bad news would take at least ten years to reach Her. And then it would be another ten years, at least, before She could get there to do anything about it...

You may answer that this is terribly naive—that God is already 'everywhere.' Perhaps so, but that really comes to the same thing as saying that Her thoughts and Her influence can travel at an infinite velocity. And in this case, the Einstein speed limit is not absolute; it *can* be broken.

The implications are profound. From the human viewpoint it is no longer absurd—though it may be presumptuous—to hope that we may one day have knowledge of the most distant parts of the universe. The snail's pace of the velocity of light need not be an eternal limitation, and the remotest galaxies may one day lie within our reach.

But perhaps, on the other hand, God Herself is limited by the same laws that govern the movements of electrons and protons, stars and spaceships. And that may be the cause of all our troubles.

She's coming just as quickly as She can, but there's nothing that even She can do about that maddening 186,000 miles a second.

It's anybody's guess whether She'll be here in time.

"Impudence—my guardian angel." Perhaps the most famous of all the photographs of Einstein. It was taken on his 72nd birthday in 1951, as he was about to leave after giving the first Einstein Awards for Achievements in Natural Sciences.

CHRONOLOGY OF EINSTEIN'S LIFE

1879 Albert Einstein born in Ulm, Germany (14 March).

1881 Sister Maria (Maja) born.

1884 Father shows him a magnetic compass.

1885 Begins school in Munich.

1888 Moves school to Luitpold Gymnasium in Munich.

1889–92 Begins private study of mathematics and science. Does not become a bar mitzvah.

1894 Leaves Luitpold Gymnasium without graduating; joins parents in Italy.

1895–96 Attends cantonal school in Aarau, Switzerland.

1896 Released from German nationality.

1896–1900 Attends Swiss Polytechnic (Zurich) and graduates.

1901 Acquires Swiss citizenship. Applies for scientific research posts without success; publishes first scientific paper.

1902 Daughter born to Mileva Marić. Begins work at Patent Office in Bern (23 June).

1903 Marries Mileva Marić; their daughter may have been adopted. Starts Olympia Academy with Maurice Solovine and Conrad Habicht.

1904 First son Hans Albert born.

1905 Writes five scientific papers (the second is his doctoral thesis), on light quanta (March), molecular dimensions (April), Brownian movement (May), special relativity (June) and $E=mc^2$ (September).

1906 Writes paper on specific heat using quantum theory.

1907 Discovers equivalence principle crucial to general relativity.

1908 Special relativity reformulated in space-time by Hermann Minkowski.

1909 Lectures at Salzburg and announces wave-particle duality. Leaves Patent Office and moves to Zurich as extraordinary professor of theoretical physics.

1910 Second son Eduard born.

1911 Moves to Prague as professor of theoretical physics. Attends first Solvay Congress in Brussels.

1912 Returns to Zurich as professor of theoretical physics.

1914 Moves to Berlin as member of the Prussian Academy of Sciences. Albert and Mileva Einstein separate; she returns to Zurich with their sons. Publicly opposes German nationalism on outbreak of war.

1915 Announces general theory of relativity (November).

1916 Writes papers on quantum theory of radiation, including concept of spontaneous and stimulated emission and absorption.

1917 Writes paper about structure of universe which introduces cosmological constant.

1918 Writes paper on gravitational waves. Welcomes collapse of German Reich in First World War.

1919 Divorces Mileva and marries cousin Elsa Einstein. British astronomers observe solar eclipse (29 May) which confirms bending of starlight by the Sun predicted by general theory of relativity.

1920 Public attacks on relativity theory and 'Jewish physics' begin in Germany.

1921 First visit to United States, on fundraising tour for Hebrew University in Jerusalem.

1922 Visit to Japan. Awarded Nobel prize for 1921.

1923 Visit to Palestine; gives first lecture at Hebrew University in Jerusalem. Publishes first attempt at unified theory of gravitation and electromagnetism.

1925 Publishes two papers on Bose-Einstein statistics and Bose-Einstein condensation.

1925–26 Quantum mechanics founded by Werner Heisenberg, Erwin Schrödinger and others; Einstein is sceptical.

1927 Attends Solvay Congress and begins dispute with Niels Bohr about quantum mechanics.

1929 Receives Max Planck Medal on Planck's seventieth birthday.

1930–33 Three visits to California Institute of Technology for research.

1933 Announces he will not return to Germany after Nazi seizure of power; resigns from Prussian Academy; moves to United States and works at Institute for Advanced Study in Princeton. Never visits Europe again.

1935 Publishes Einstein-Podolsky-Rosen paradox about quantum mechanics.

1936 Second wife Elsa dies in Princeton.

1938 Publishes *The Evolution of Physics* with Leopold Infeld. Son Hans Albert and family emigrate to United States.

1939 Sister Maja joins him in Princeton. Signs letter to President Roosevelt urging development of atomic bomb against Germany.

1940 Acquires American citizenship (retains Swiss citizenship).

1943 Begins war work for US Navy but is not involved in atomic bomb project.

1946 Assumes chairmanship of Emergency Committee of Atomic Scientists. Publicly champions need for arms control and world government. Also publicly opposes racism in United States.

1948 Diagnosed with life-threatening aneurysm. His first wife Mileva dies in Zurich.

1950 Stipulates in will that the final depository of his papers be the Hebrew University in Jerusalem. Publicly opposes building of hydrogen bomb. Secret inquiry into Einstein as potential Communist subversive ordered by FBI director J. Edgar Hoover.

1951 Sister Maja dies in Princeton.
1952 Offered presidency of Israel but declines.
1953–54 Publicly opposes McCarthyism, creating fierce controversy.
1955 Signs Russell-Einstein manifesto against spread of nuclear weapons. Dies in Princeton (18 April), still working on unified theory.

SOURCES OF QUOTATIONS

Preface (Dyson)

9 **The essential result of this investigation** Albert Einstein, "Stationary system with spherical symmetry consisting of many gravitating masses," *Annals of Mathematics*, (Series 2) 40, 1939: 936.

1 Physics before Einstein

12 **In one person [Newton] combined** Foreword to Isaac Newton, *Opticks*, London: Bell, 1931: vii.
12 **The whole of science** "Physics and reality" in Einstein, *Ideas and Opinions*: 290.
13 **The moving body...push it** Einstein and Infeld: 7.
13 **All the philosophy of nature** Gleick: 52.
13 **taught man to be modest** "Message on the 410th anniversary of the death of Copernicus" in Einstein, *Ideas and Opinions*: 359.
14 **It seems that the human mind** "Johannes Kepler" in ibid: 266.
14 **Pure logical thinking** "On the method of theoretical physics" in ibid: 271.
15 **Shut yourself up** Giulini: 12–13.
16 **the greatest bodies of the universe** Gleick: 189.
16 **How does the state of motion** "The mechanics of Newton and their influence on the development of theoretical physics" in Einstein, *Ideas and Opinions*: 255.
17 **Every body perseveres** Newton (Cohen and Whitman): 416–17.
17 **A change in motion is proportional** Ibid.
19 **To any action there is always** Ibid.
21 **It is enough that gravity** Ibid: 943.
21 **Absolute, true, and mathematical time** Ibid: 408.
21 **God informed** Gleick: 152.
21 **It may be, that there is no** Newton (Motte): 8.
21 **eminently fruitful** "The fundaments of theoretical physics" in Einstein, *Ideas and Opinions*: 325.
22 **He was justified in sticking** Ibid: 326.
24 **absolutely stationary** Kaku: 11.
25 **Before Maxwell** "Maxwell's influence on the evolution of the idea of physical reality" in Einstein, *Ideas and Opinions*: 269.

2 The Making of a Physicist

32 **For the detective** Einstein and Infeld: 78.
32 **exploration of my ancestors** Letter to Hans Mühsam (4 Mar. 1953) in Seelig: 56.
32 **entirely irreligious** Schilpp: 3.
33 **Let us return to Nature** Moszkowski: 66.
35 **lie down on the sofa** Einstein, *Collected Papers*, 1: lxiv.
35 **Einstein was more** Gerald Whitrow in Whitrow: 52.
35 **Einstein expressed** Born and Einstein: 105.
35 **I can still remember** Schilpp: 9.
35 **a second wonder** Ibid.
35 **earth measuring** "Geometry and experience" in Einstein, *Ideas and Opinions*: 234.
36 **suspicion against every kind of authority** Schilpp: 5.
36 **he would never get anywhere** Einstein, *Collected Papers*, 1: lxiii.
36 **To punish me** Hoffmann: 24.
37 **a personal gift** Ibid: 28.
38 **creating a new theory** Einstein and Infeld: 159.
39 **I'm convinced more and more** Einstein and Marić (10? Aug. 1899): 10.
39 **impudence** Ibid (12 Dec. 1901): 67.
40 **I will have soon graced** Ibid (4 Apr. 1901): 42.

A Brief History of Relativity (Hawking)

51 **Politics is for the present** Seelig: 71.

3 The Miraculous Year, 1905

52 **The eternal mystery of the world** "Physics and reality" in Einstein, *Ideas and Opinions*: 292.
52 **It deals with radiation** Fölsing: 120.
53 **was intrigued rather than dismayed** Rigden: 8.
56 **inventions of the intellect** "Johannes Kepler" in Einstein, *Ideas and Opinions*: 266.
56 **[Einstein] would carefully study** Jürgen Renn and Robert Schulmann in introduction to Einstein and Marić: xxii.
56 **These investigations of Einstein** Schilpp: 166.
59 **three intellectual musketeers** Highfield and Carter: 96.
59 **So that was caviar** Ibid: 98.
59 **laughed so much** Ibid: 102.
59 **far less childish** Ibid: 97.
60 **Thank you. I've completely solved** Fölsing: 155.
60 **steadfastness** Ibid: 195.
61 **misdeed** Einstein, *Relativity*: 10.

61 **The stone traverses** Ibid: 11.
61 **If I pursue a beam of light** Schilpp: 53.
62 **We should catch them** Einstein and Infeld: 177.
62 **unjustifiable hypotheses** Einstein, *Relativity*: 32.
62 **Not only do we have** Fölsing: 175.
62 **We are accustomed** "Physics and reality" in Einstein, *Ideas and Opinions*: 299.
63 **Is the world to be described** Rigden: 21.
64 **According to the assumption** Ibid: 19.
65 **I just read a wonderful paper** Einstein and Marić (28? May 1901): 54.

4 General Relativity

66 **The years of anxious searching** "Notes on the origin of the general theory of relativity" in Einstein, *Ideas and Opinions*: 289–90.
66 **modification** Fölsing: 120.
66 **In boldness** Ibid: 271.
69 **not compatible with** Bernstein: 82.
69 **I am at my wits' end** Fölsing: 205.
69 **in my opinion, these [other] theories** Pais, '*Subtle is the Lord*': 159.
70 **Subtle is the Lord** Ibid: 113.
70 **I could not believe he could be the father** Miller: 213.
70 **The views of space and time** Bernstein: 95.
71 **Since the mathematicians** Fölsing: 245.
71 **as far as the propositions of mathematics** "Geometry and experience" in Einstein, *Ideas and Opinions*: 233.
71 **mysterious shuddering** Einstein, *Relativity*: 56.
72 **The two sentences** Einstein and Infeld: 224.
72 **I was sitting on a chair** Fölsing: 301.
72 **the happiest thought** Miller: 217.
73 **the idea that the physics in an accelerated laboratory** Hey and Walters, *Einstein's Mirror*: 270.
73 **You understand, what I need to know** Fölsing: 325.
74 **A beam of light carries** Einstein and Infeld: 234.
75 **When a blind beetle** Quoted in Michael Grüning, *Ein Haus für Albert Einstein*, Berlin: Verlag der Nation, 1990: 498.
75 **space was deprived of its rigidity** "The problem of space, ether, and the field in physics" in Einstein, *Ideas and Opinions*: 281.
76 **The examination of the correctness** Einstein, *Relativity*: 76-77.
76 **I knew all the time** Fölsing: 439.
76 **But, you know, [Planck] didn't really understand** French: 31.
77 **had he been left to himself** Chandrasekhar: 112.

5 Arguing about Quantum Theory

82 **Quantum mechanics is certainly imposing** Born and Einstein (4 Dec. 1926): 88.
82 **I think I can safely say** Hey and Walters, *The New Quantum Universe*: 1.
83 **services to theoretical physics** Pais, '*Subtle is the Lord*': 386.
83 **it was the *law* that was accepted** Whitaker: 100.
83 **Einstein's initial conception of the photon** Ibid: 125.
84 **the tremendous practical** "On the method of theoretical physics" in Einstein, *Ideas and Opinions*: 273.
84 **probably the strangest thing** Fölsing: 154.
85 **That sometimes, as for instance** Ibid: 147.
85 **wholly untenable** Pais, '*Subtle is the Lord*': 357.
87 **If Planck's theory of radiation** Ibid: 395.
88 **It cannot be denied** Fölsing: 256.
88 **It is my opinion that the next phase** Rigden: 37–38.
88 **one of the landmarks** Schilpp: 154.
89 **revelation** Fölsing: 390.
89 **Such questions plagued the theory** Whitaker: 115.
90 **Einstein was the first to recognize** Townes: 13.
90 **With this, the light quanta** Fölsing: 392.
90 **leaves the time and direction** Ibid: 392.
92 **Were it not for Einstein's challenge** Jammer: 220.
92 **if one abandons** Born and Einstein: 162.
92 **Einstein's moon really exists** Peat: 166.
93 **It was quite a shock for Bohr** Whitaker: 217–18.
93 **wave, or quantum, mechanics** Pais, '*Subtle is the Lord*': 515.
94 **The conviction prevails** Einstein, *Relativity*: 158.

6 The Search for a Theory of Everything

95 **No fairer destiny** Einstein, *Relativity*: 78.
95 **He saw in the quantum mechanics of today** Born and Einstein: 199.
95 **to obtain a formula** Jammer: 57.
96 **Could we not reject** Einstein and Infeld: 257–58.
96 **no good** Fölsing: 563.
96 **Our experience hitherto** "On the method of theoretical physics" in Einstein, *Ideas and Opinions*: 274.
98 **completely cuckoo** Fölsing: 693.
98 **alchemist** Magueijo: 235.
98 **would be undiminished** Pais, *Einstein Lived Here*: 43.
98 **highly esteemed** Fölsing: 692.
98 **One feels as if one were an Ichthyosaurus** Born and Einstein (12 May 1952): 188.
98 **He himself had established his name** Abraham Taub quoted by Christopher Sykes in foreword to Whitrow: xii.

98 **There are two different conceptions** Tagore: 531–32.
100 **we ought to be concerned solely** Heisenberg: 68.
100 **wrong to think** Pais, *Niels Bohr's Times*: 427.
100 **Man defends himself** Tagore: 532.
100 **I have never been able to understand Einstein** Born and Einstein: 151.

Einstein's Search for Unification (Weinberg)

107 **all attempts to obtain** "On the generalized theory of gravitation" in Einstein, *Ideas and Opinions*: 352.

7 Physics since Einstein

109 **Science is not and never will** Einstein and Infeld: 308.
111 **to be reformulated** Ed Fomalont and Sergei Kopeikin, "How fast is gravity?," *New Scientist*, 11 Jan. 2003: 33.
113 **scores of manuscripts** Rigden: 98.
115 **necessary only for the purpose** Singh: 148.
115 **However the possibility of a cosmological constant** Weinberg: 178–79.
117 **It was clear that anything that falls** Thorne: 121.
118 **A hundred years later technologists** *Scientific American*, Sept. 2004: 29.
119 **Imagine an atom in the gas** Hey and Walters, *The New Quantum Universe*: 145.
119 **What would you say to the following situation?** Rigden: 145.
120 **on one supposition we should** Schilpp: 85.
120 **spooky actions at a distance** Born and Einstein (3 Mar. 1947): 155.
120 **Science cannot solve** Kaku: 162.

Einstein's Scientific Legacy (Anderson)

122 **a sound mathematical education** Schilpp: 15.
122 **during my student years** Seelig: 11.

8 The Most Famous Man in the World

130 **I never understood why** Preface written in 1942 but not published until 1979 in Philipp Frank, *Einstein: Sein Leben und seine Zeit*, in Fölsing: 457.
131 **like the hermits of old** Letter to Michele Besso (5 Jan. 1929) in Fölsing: 604.
131 **Large crowds gather** Pais, *Einstein Lived Here*: 179.
131 **They cheer me** Fölsing: 457.
131 **The speed with which his fame** Clark: 246.
132 **The whole atmosphere of tense** Bernstein: 119.
132 **A very definite result** Fölsing: 443.
132 **this result is not an isolated one** Chandrasekhar: 116.
133 **Professor Eddington, you must be** Ibid: 117.
134 **The supposed astronomical proofs** Brian: 101–02.
134 **the moronic brain child** Ibid: 103.
134 **science riot** Singh: 143.
134 **Einstein must never receive a Nobel** Rigden: 100.
134 **Einstein believes his books** Fölsing: 379.
134 **An hour sitting with a pretty girl** Sayen: 130.
136 **We all know that no iron curtain** Moszkowski: 72.
136 **For scientific relations** Fölsing: 445.
136 **The mere thought that a living** Moszkowski: 14.
136 **The term 'theory of relativity'** Jammer: 33–34.
137 **It meant the dethronement of time** Ibid: 159.
139 **a kind of great man** Pais, *Einstein Lived Here*: 181.
139 **the approximately 3000 participants** Fölsing: 527.
139 **Why is it that nobody understands me** *New York Times*, 12 Mar. 1944.
139 **I really cannot understand why** Born and Einstein (12 Apr. 1949): 179.

9 Personal and Family Life

140 **What I most admired** Letter to Vero and Bice Besso (21 Mar. 1955) in Pais, *Einstein Lived Here*: 25.
140 **He left this world** Born and Einstein: 229.
140 **I am truly a 'lone traveller'** "The world as I see it" in Einstein, *Ideas and Opinions*: 9.
141 **Nothing tragic really gets to him** Letter to Antonina Vallentin (26 Oct. 1934) in Fölsing: 685.
141 **When I'm not with you** Einstein and Marić (6 Aug. 1900): 23–24.
141 **You are and will remain** Ibid (27 Mar. 1901): 39.
141 **When you're my dear little wife** Ibid (28 Dec. 1901): 72–73.
141 **human happiness** Mileva Marić to Helene Savić (Dec.? 1901) in Marić: 79.
142 **calmly philosophic** Highfield and Carter: 130.
142 **He was sitting in his study** Ibid.
142 **the papers he has written** Mileva Marić to Helene Savić (Dec. 1906?) in Marić: 88.
142 **He is now regarded as the best** Mileva Marić to Helene Savić (3 Sept. 1909) in ibid: 98.
143 **Very few women are creative** Quoted by Esther Salaman in a BBC radio talk, *Listener*, 8 Sept. 1955.
143 **very intelligent but has the soul of a herring** Highfield and Carter: 158.
143 **strength of character and devotion** "Marie Curie in Memoriam" in Einstein, *Ideas and Opinions*: 77.
143 **I'm not much with people** Quoted by Esther Salaman in a BBC radio talk, *Listener*, 8 Sept. 1955.
143 **Einstein was a flirt** Highfield and Carter: 52.
144 **She might have passed in her prime** Ibid: 145.
144 **I think it's a disgusting idea** Michelmore: 79.

146 **secretory causes** Letter to Michele Besso (21 Oct. 1932) in Fölsing: 673.
146 **He represents virtually the only human problem** Letter to Carl Seelig (20 Apr. 1952) in ibid: 731.
147 **There is a block behind it** Letter to Carl Seelig (4 Jan. 1954) in ibid: 731.
147 **Strangely enough, her intelligence** Letter to Lina Kocherthaler (27 July 1951) in ibid: 731.

Einstein's Love Letters (Schulmann)
151 **jealous of science** Mileva Einstein to Helene Savić. See Marić: 102.
151 **strenuous intellectual work** Einstein, *Collected Papers*, 1: 55–56.
152 **passionate interest** Einstein, *Mein Weltbild*: 10. See also Einstein, "The world as I see it," *Ideas and Opinions*: 9.

Einstein and Music (Glass)
153 **If I were not a physicist** Interview with G. S. Viereck, "What life means to Einstein," *Saturday Evening Post*, 26 Oct. 1929.
153 **excellent** Einstein, *The New Quotable Einstein*: 330.

10 Germany, War and Pacifism
156 **A funny lot** Written on 17 Apr. 1925, in Fölsing: 549.
156 **That a man can take pleasure** "The world as I see it" in Einstein, *Ideas and Opinions*: 10–11.
158 **psychoses** Einstein and Sigmund Freud, *Why War?*, Paris: International Institute of Intellectual Co-operation, 1933: 18.
158 **the lies and defamations** Fölsing: 345.
158 **create an organic unity** Stern: 115.
160 **Why are we hated** Fölsing: 366.
160 **He confines himself** Ibid: 349.
160 **Most physicists were...part of the war effort** Ibid: 398.
160 **In Haber** L. F. Haber, *The Poisonous Cloud: Chemical Warfare in the First World War*, Oxford: 1986: 27.
162 **The best minds from all epochs** Fölsing: 367–68.
163 **I am reminded of the period of the witch** Ibid: 399.
163 **tantamount to a vile breach** Ibid: 431.
163 **He invented the difference** Born and Einstein: 35.
163 **In contrast to the intractable** Calaprice: 14.
163 **a wholly irreversible accumulation** Born and Einstein (12 Feb. 1921): 52.
164 **If even two percent** Fölsing: 635.
164 **Turn around** Pais, *Einstein Lived Here*: 190.
164 **But now the war of extermination** Fölsing: 664.
164 **Two ideologies** Ibid.

11 America
166 **We are unjust** "Some impressions of the USA" in Einstein, *Ideas and Opinions*: 7 (which incorrectly attributes this piece to 1921).
166 **If it is successful, radioactive poisoning** "National security" in ibid: 159–60.
168 **derogatory information** Jerome: 156.
169 **No one ever accused Hoover** Ibid: xxi.
169 **Making allowances** Ibid: 35.
170 **Naturally, I am needed** Fölsing: 495.
170 **because of the large percentage** Cassidy: 121.
170 **Concern for man himself** Fölsing: 640.
171 **It is an illegal fight** Jerome: 23.
171 **My hope is that your visit** Fölsing: 642.
172 **Not even Stalin himself** Jerome: 7.
172 **But are they not perfectly right** Ibid: 9.
173 **nearly opposites in temperament** Cassidy: 346.
174 **I can only say I have the greatest respect** Pais, *Einstein Lived Here*: 241.
174 **The reactionary politicians** "Modern inquisitional methods" in Einstein, *Ideas and Opinions*, 33–34.
175 **anyone who gives advice like Einstein's** Pais, *Einstein Lived Here*: 238.
175 **That's the same advice** Jerome: 240.
175 **a disloyal American** Ibid.
175 **Americans are proud** Ibid: 218.
175 **No other man contributed so much** *New York Times*, 19 Apr. 1955.

12 Zionism, the Holocaust and Israel
176 **I am neither a German citizen** Clark: 379.
177 **About God, I cannot accept** Jammer: 122.
177 **There is nothing divine about morality** "The religious spirit of science" in Einstein, *Ideas and Opinions*: 40.
177 **There is only one fault** Clark: 426.
177 **Edgar, I'm used to pomp** Jerome: 223.
178 **In the philosophical sense** "Is there a Jewish point of view?" in Einstein, *Ideas and Opinions*: 185–86.
179 **It goes against the grain** Fölsing: 489.
179 **you know that my darling** Letter to Helene Savić (Dec.? 1901) in Marić: 79.
179 **Herr Dr Einstein** Fölsing: 250.
179 **Why are these fellows** Ibid: 489–90.
179 **I am delighted to know** Ibid: 668.
179 **was Jewish, but he wished he weren't** Isidor Rabi quoted in Cassidy: 32.
179 **unjustified humiliations** Born and Einstein: 17.
179 **not to be got rid of** "Addresses on reconstruction in Palestine" (V) in Einstein, *Ideas and Opinions*: 182.

180 **History has shown that Einstein** Born and Einstein: 17.
181 **a complete liar** Fölsing: 671.
181 **acquiring the psychology** Ibid: 671.
181 **dull-minded tribal companions** Ibid: 529.
181 **My heart says yes** Ibid: 532.
181 **a nasty mess** Born and Einstein (30 May 1933): 111.
182 **ambitious and weak person** Fölsing: 594.
182 **If the Nazis claimed** Jerome: 26.
182 **The Germans as an entire people** "To the heroes of the battle of the Warsaw Ghetto" in Einstein, *Ideas and Opinions*: 212–13.
182 **The attitude of the German intellectuals** Fölsing: 728.
182 **migrating back** Born and Einstein (12 Oct. 1953): 195.
182 **The hours which I was permitted** Fölsing: 729.
184 **the essential nature of Judaism** "Our debt to Zionism" in Einstein, *Ideas and Opinions*: 190.
184 **I am deeply moved** Fölsing: 733.
184 **Tell me what to do** Ibid: 734.

Einstein on Religion, Judaism and Zionism (Jammer)
185 **disguised theologian** Dürrenmatt: 12.
185 **deeply religious nonbeliever** Letter to Hans Mühsam (30 Mar. 1954) in Albert Einstein Archives: 38: 434.
185 **religious paradise of youth...lies** Schilpp: 5.
186 **everyone who is seriously** Letter to Phyllis Wright (24 Jan. 1936) in Albert Einstein Archives: 42: 601.
186 **cosmic religious feeling...scientific research** "Religion and science" in Einstein, *Ideas and Opinions*: 39.
186 **science without religion is lame** "Science and religion" in ibid: 46.
186 **Einstein did not think that religious faith** Born and Einstein: 199.
186 **in view of such a harmony** Hubertus zu Löwenstein, *Towards the Further Shore: An Autobiography*, London: Victor Gollancz, 1968: 156.
186 **Now I know there is a God** Brian: 193.
188 **By an application of the theory of relativity** *The Times*, 28 Nov. 1919.
188 **What has that to do with Zionism?** Clark: 378.
188 **I am neither a German citizen** Ibid: 379.
188 **I thank you** Letter to Blumenfeld (25 Mar. 1955) in Albert Einstein Archives: 59: 274.
188 **It was in America** *Jüdische Rundschau*, 1 July 1921. See "On a Jewish Palestine" in Einstein, *Collected Papers*, 7.
189 **Mount the platform** Clark: 394.
189 **I have already had the privilege** Ibid: 394.
189 **I know a little about nature** Ibid: 618.
189 **the greatest thing in Palestine** Ibid: 388.
189 **never cease to regard the fate** Fölsing: 595.
189 **a great spiritual centre** Message on the opening of the Hebrew University in 1925 reprinted in Rosenkranz: 97.
189 **Should we be unable** Letter to Chaim Weizmann (25 Nov. 1929) in Albert Einstein Archives: 33: 411.
190 **welfare of the whole population** "Letter to an Arab" (15 Mar. 1930) in Einstein, *Ideas and Opinions*: 173.
190 **One who, like myself** Letter to *Falastin* (Dec. 1929), published on 28? Jan. 1930.
190 **When appraising the achievement** "The Jews of Israel" in Einstein, *Ideas and Opinions*: 201.

13 Nuclear Saint and Demon
191 **The feeling for what ought** Born and Einstein (7 Sept. 1944): 145.
192 **It is possible that all heavy matter** Friedman and Donley: 164.
192 **The nation which can transmute** Frederick Soddy, *The Interpretation of Radium*, London: John Murray, 1909: 244.
193 **At present there is not the slightest indication** Moszkowski: 24.
193 **talking moonshine** Hey and Walters, *The New Quantum Universe*: 288.
193 **in a neighbourhood that has few birds** Fölsing: 709.
193 **The results gained thus far** Pais, *Einstein Lived Here*: 216.
195 **I never thought of that!** Rhodes: 305.
195 **This new phenomenon** Rosenkranz: 74.
196 ***Oj weh*** Pais, *Einstein Lived Here*: 219.
196 **It is consistent with what we know** Dyson: 98.
196 **The war is won, but the peace is not** Einstein, *Ideas and Opinions*: 115–17.
196 **commingled and distributed** "Atomic war or peace" in ibid: 130.
196 **Do I fear the tyranny** Ibid: 120.
200 **the authority of the General Assembly** Pais, *Einstein Lived Here*: 234.
200 **By an irony of fate, Einstein** "Exchange of letters with members of the Russian Academy" in Einstein, *Ideas and Opinions*: 139.
200 **If we hold fast to the concept** Ibid: 146.
200 **tough, lucid** Bernstein: 182.
200 **If Einstein's ideas** *Time*, 31 Dec. 1999: 37.
200 **We really should not be surprised** Born and Einstein (7 Sept. 1944): 145.

Einstein's Quest for Global Peace (Rotblat)
202 **My pacifism...hatred** Nathan and Norden: 98.
202 **To me the killing** Ibid: 93.

203 **War constitutes** Ibid: 54.
203 **Were it not for German militarism** Ibid: 3.
205 **Had I known that the Germans** Fölsing: 725.
205 **Thank you for your letter of April 5** Nathan and Norden: 631.
206 **There lies before us** Ibid: 635.

14 The End of an Era
207 **Is there not a certain satisfaction** Albert Einstein Archives: 5: 150.
207 **Now he has departed** Letter to Vero and Bice Besso (21 Mar. 1955) in Einstein, *The New Quotable Einstein*: 73.
207 **The great scientist of our age** Ibid: 315.
207 **A powerful searchlight** Ibid: 322.
207 **the brightest jewel** Ibid: 326.
207 **He was one of the greats of all ages** Ibid: 316.
207 **Of all the public figures** Ibid: 320–21.
208 **Through Albert Einstein's work** Pais, *Einstein Lived Here*: 255–56.
208 **played major roles** Ibid: 256.
208 **I loved him and admired him** Einstein, *The New Quotable Einstein*: 324.
209 **hero worshippers and cranks** H. Fleming quoted in Highfield and Carter: 260.
210 **Probably the only project** Pais, *Einstein Lived Here*: 199.
210 **I want go when *I* want** Pais, *'Subtle is the Lord'*: 477.
210 **as in Haydn's *Farewell Symphony*** Letter to Boris Schwarz (1945) in Albert Einstein Archives: 79: 678.
210 **completely in command** Born and Einstein: 229.
210 **worshipped** Quoted by Abraham Pais in *Manchester Guardian*, 17 Dec. 1994.

Einstein's Last Interview (Cohen)
213 **Newton, forgive me** Schilpp: 31.
218 **it alone can afford us** Foreword to Isaac Newton, *Opticks*, London: Bell, 1931: viii.
220 **Newton,... you found** Schilpp: 31.

15 Einstein's Enduring Magic
226 **Knowledge exists in two** "Message in honour of Morris Raphael Cohen" in Einstein, *Ideas and Opinions*: 80.
226 **a place of pilgrimage** Einstein, *The New Quotable Einstein*: 62.
226 **reveal the secrets of genius** Review of Carolyn Abraham's *Possessing Genius* in *Times Higher Education Supplement*, 1 Oct. 2004.
226 **unworthy to lace** *Times Higher Education Supplement*, 29 Oct. 2004.
226 **Things should be made as simple** Various slightly different versions of this quotation are attributed to Einstein but there is no definitive source. See the discussion in Einstein, *The New Quotable Einstein*: 290–91.
228 **...If cloning** Friedman and Donley: 188.
228 **What did this mega-genius eat?** *Scientific American*, Mar. 2003: 84.
228 **Whenever we came to an impasse** Whitrow: 75.
229 **crackpot missives** *Scientific American*, Sept. 2004: 80.
230 **He was of the most fearful** Gleick: 228.
230 **Newton is the Old Testament god** Rigden: 149–50.
230 **Blush, Born, Blush!** Born and Einstein: 161.
231 **the greatest political genius** Nathan and Norden: 584.
231 **Einstein was profoundly spiritual** *Times Higher Education Supplement*, 29 Oct. 2004.
231 **It would be perfectly consistent** Jammer: 264.
231 **the mind of God** Stephen Hawking, *A Brief History of Time*, London: Bantam, 1988: 174.
231 **Filled with admiration** Jammer: 253.
231 **time and again filled me** Fölsing: 283.
232 **Einstein of the new theatrical form** Friedman and Donley: 176.
232 **Simply because writers say** Ibid: 87.
232 **the less they know** Quoted by Esther Salaman in a BBC radio talk, *Listener*, 8 Sept. 1955.
232 **In science one tries to tell people** Attributed to Dirac, possibly said to Robert Oppenheimer in 1927.
232 **Mozart's music** Quoted in Armin Hermann, *Albert Einstein*, Munich: Piper, 1994: 158.

Einstein: Twentieth-Century Icon (Clarke)
236 **I do not know how the Third World War** Interview with Alfred Werner, *Liberal Judaism*, Apr.-May 1949: 12.

BIBLIOGRAPHY

Abraham, Carolyn, *Possessing Genius: The Bizarre Odyssey of Einstein's Brain*, Cambridge UK: Icon, 2004

Bernstein, Jeremy, *Einstein*, London: Fontana Modern Masters, 2nd edn, 1991

Bodanis, David, *E=mc²: A Biography of the World's Most Famous Equation*, London: Macmillan, 2000

Born, Max and Albert Einstein, *The Born-Einstein Letters*, London: Macmillan, 2nd edn, 2005

Brian, Denis, *Einstein: A Life*, New York: Wiley, 1996

Calaprice, Alice, *The Einstein Almanac*, Baltimore: Johns Hopkins University Press, 2005

Cassidy, David C., *J. Robert Oppenheimer and the American Century*, New York: Pi Press, 2004

Chandrasekhar, Subrahmanyan, *Truth and Beauty: Aesthetics and Motivations in Science*, Chicago: Chicago University Press, 1987

Clark, Ronald W., *Einstein: The Life and Times*, New York: World, 1971

Collins, Harry, *Gravity's Shadow: The Search for Gravitational Waves*, Chicago: Chicago University Press, 2004

Dürrenmatt, Friedrich, *Albert Einstein: Ein Vortrag*, Zurich: Diogenes Verlag, 1979

Dyson, Freeman, *Imagined Worlds*, Cambridge MA: Harvard University Press, 1997

Einstein, Albert:

The Collected Papers of Albert Einstein, Vols 1–9, various editors, Princeton: Princeton University Press, 1987–

Ideas and Opinions, Carl Seelig, ed., New York: Three Rivers Press, 1982

Mein Weltbild, Amsterdam: Querido Verlag, 1934

The New Quotable Einstein, Alice Calaprice, ed., Princeton: Princeton University Press, 2005

Relativity: The Special and the General Theory, London: Routledge Classics, 2001

Einstein, Albert and Leopold Infeld, *The Evolution of Physics: The Growth of Ideas from the Early Concepts to Relativity and Quanta*, Cambridge UK: Cambridge University Press, 1938

Einstein, Albert and Mileva Marić, *Albert Einstein/Mileva Marić: The Love Letters*, Jürgen Renn and Robert Schulmann, eds, Princeton: Princeton University Press, 1992

Fölsing, Albrecht, *Albert Einstein: A Biography*, London: Viking, 1997

French, A. P., ed., *Einstein: A Centenary Volume*, London: Heinemann, 1979

Friedman, Alan J. and Carol C. Donley, *Einstein as Myth and Muse*, Cambridge UK: Cambridge University Press, 1985

Giulini, Domenico, *Special Relativity: A First Encounter*, Oxford: Oxford University Press, 2005

Gleick, James, *Isaac Newton*, London: Fourth Estate, 2003

Hawking, Stephen, *The Universe in a Nutshell*, London: Bantam, 2001

Heisenberg, Werner, *Physics and Beyond: Encounters and Conversations*, London: Allen and Unwin, 1971

Hey, Tony and Patrick Walters:

Einstein's Mirror, Cambridge UK: Cambridge University Press, 1997

The New Quantum Universe, Cambridge UK: Cambridge University Press, 2003

Highfield, Roger and Paul Carter, *The Private Lives of Albert Einstein*, London: Faber and Faber, 1993

Hoffmann, Banesh, *Albert Einstein: Creator and Rebel*, New York: Viking, 1972

Jammer, Max, *Einstein and Religion: Physics and Theology*, Princeton: Princeton University Press, 1999

Jerome, Fred, *The Einstein File: J. Edgar Hoover's Secret War Against the World's Most Famous Scientist*, New York: St Martin's Press, 2002

Kaku, Michio, *Einstein's Cosmos: How Albert Einstein's Vision Transformed Our Understanding of Space and Time*, London: Weidenfeld and Nicolson, 2004

Magueijo, João, *Faster Than the Speed of Light: The Story of a Scientific Speculation*, London: Arrow, pbk edn, 2004

Marić, Mileva, *The Life and Letters of Mileva Marić, Einstein's First Wife*, Milan Popović, ed., Baltimore: Johns Hopkins University Press, 2003

Michelmore, Peter, *Einstein: Profile of the Man*, New York: Dodd, Mead, 1962

Miller, Arthur I., *Einstein, Picasso: Space, Time and the Beauty That Causes Havoc*, New York: Basic Books, 2001

Moszkowski, Alexander, *Conversations with Einstein*, London: Sidgwick and Jackson, 1972

Nathan, Otto and Heinz Norden, eds, *Einstein on Peace*, New York: Schocken, 1960

Newton, Isaac:

The Principia: Mathematical Principles of Natural Philosophy, I. Bernard Cohen and Anne Whitman, trans, Berkeley: University of California Press, 1999

Sir Isaac Newton's Mathematical Principles of Natural Philosophy and His System of the World, Andrew Motte, trans. (1729), revised by Florian Cajori, Berkeley: University of California Press, 1947

Pais, Abraham:

Einstein Lived Here, New York: Oxford University Press, 1994

Niels Bohr's Times in Physics, Philosophy, and Polity, New York: Oxford University Press, 1991

'Subtle is the Lord': The Science and Life of Albert Einstein, New York: Oxford University Press, pbk edn, 1983

Peat, F. David, *Einstein's Moon: Bell's Theorem and the Curious Quest for Quantum Reality*, Chicago: Contemporary, 1990

Rhodes, Richard, *The Making of the Atomic Bomb*, New York: Simon and Schuster, 1986

Rigden, John S., *Einstein 1905: The Standard of Greatness*, Cambridge MA: Harvard University Press, 2005

Rosenkranz, Ze'ev, *The Einstein Scrapbook*, Baltimore: Johns Hopkins University Press, 2002

Rosenkranz, Ze'ev, ed., *Albert through the Looking Glass*, Jerusalem: Jewish National and University Library, 1998

Sayen, Jamie, *Einstein in America: The Scientist's Conscience in the Age of Hitler and Hiroshima*, New York: Crown, 1985

Schilpp, Paul Arthur, ed., *Albert Einstein: Philosopher-Scientist*, Evanston: The Library of Living Philosophers, 1949 (includes Einstein's "Autobiographical Notes")

Scientific American, "Beyond Einstein" (special issue on Einstein), September 2004

Seelig, Carl, ed., *Helle Zeit, Dunkle Zeit: In Memoriam Albert Einstein*, Zurich: Europa Verlag, 1956

Singh, Simon, *Big Bang*, London: Fourth Estate, 2004

Stern, Fritz, *Einstein's German World*, London: Penguin, pbk edn, 2001

Tagore, Rabindranath, *Selected Letters of Rabindranath Tagore*, Krishna Dutta and Andrew Robinson, eds, Cambridge UK: Cambridge University Press, 1997

Thorne, Kip S., *Black Holes and Time Warps: Einstein's Outrageous Legacy*, London: Macmillan, pbk edn, 1995

Townes, Charles H., *How the Laser Happened: Adventures of a Scientist*, New York: Oxford University Press, 1999

Weinberg, Steven, *Dreams of a Final Theory: The Search for the Fundamental Laws of Nature*, London: Hutchinson Radius, 1993

Whitaker, Andrew, *Einstein, Bohr and the Quantum Dilemma*, Cambridge UK: Cambridge University Press, 1996

Whitrow, G. J., ed., *Einstein: The Man and His Achievement*, New York: Dover, 1967

ACKNOWLEDGEMENTS

Andrew Robinson

I would like to thank: Oliver Craske for conceiving the book and believing in it; Freeman Dyson, Stephen Hawking, João Magueijo, Steven Weinberg, Philip Anderson, Robert Schulmann, Philip Glass, Max Jammer, Joseph Rotblat and Arthur C. Clarke for their expert contributions; and Colin and Pam Webb of Palazzo Editions for publishing the book.

Tony Hey, the author of two imaginative books for the general reader on relativity and quantum theory, provided me with valuable comments on the scientific aspects. Alice Calaprice (editor of *The New Quotable Einstein*), Robert Schulmann and especially Barbara Wolff (of the Albert Einstein Archives in Jerusalem) kindly checked various points about Einstein and quotations from him and other writers. Dipli Saikia sustained me throughout the research and writing in every possible way.

As for the subject of the book himself, I can do no better than quote J. Robert Oppenheimer's remark about Einstein made in 1965: "It is not too soon to start to dispel the clouds of myth and to see the great mountain peak that these clouds hide. As always, the myth has its charms, but the truth is far more beautiful."

Palazzo Editions

In addition, Palazzo Editions and Oliver Craske would like to thank: Roni Grosz, Chaya Becker, Barbara Wolff and Anat Eban at the Albert Einstein Archives for their steadfast support for the book; David Costa and Nadine Levy for their elegant design; Sally Nicholls for invaluable support with picture research; David Woodroffe for the eight diagrams; and John Woodruff for proof-reading the book. The index was compiled by Andrew Robinson. Special thanks are due to Robert Ingpen, who suggested that Einstein's last interview would be ideal for the book, and Conal Urquhart for his hospitality in Jerusalem.

CONTRIBUTORS

Andrew Robinson, the author and editor of the book, is a King's Scholar of Eton College and holds degrees from Oxford University (in science) and the School of Oriental and African Studies, London. He is the author of more than a dozen books, for both general and academic readers. These include *The Shape of the World*, *Earthshock*, *The Story of Writing* and *Lost Languages*, and definitive biographies of Satyajit Ray, *The Inner Eye*, and Rabindranath Tagore, *The Myriad-Minded Man* (the latter written with Krishna Dutta). Since 1994, he has been the literary editor of *The Times Higher Education Supplement* in London.

Freeman Dyson has been, since 1953, a physicist at the Institute for Advanced Study in Princeton, where Albert Einstein worked from 1933 to 1955. He is the author of many scientific books for the general reader, including *Disturbing the Universe*, *Infinite in All Directions* and *Imagined Worlds*. He wrote the foreword for *The Quotable Einstein*.

Stephen Hawking is the Lucasian professor of mathematics at the University of Cambridge, a chair once occupied by Sir Isaac Newton. His books for the general reader are *A Brief History of Time*, *Black Holes and Baby Universes and Other Essays* and *The Universe in a Nutshell*.

João Magueijo is reader in theoretical physics at Imperial College, London. He is the author of *Faster Than the Speed of Light: The Story of a Scientific Speculation*.

Steven Weinberg is Josey Regental professor of science at the University of Texas in Austin. He was awarded a Nobel prize in 1979 and a National Medal of Science for his work in particle physics. His books for the general reader include *The First Three Minutes*, *The Discovery of Subatomic Particles* and *Dreams of a Final Theory*.

Philip Anderson is Joseph Henry professor of physics emeritus at Princeton University. He was awarded a Nobel prize in 1977 and a National Medal of Science for his work in solid-state physics.

Robert Schulmann is a former director of the Einstein Papers Project, who continues to act as consultant to the *Collected Papers of Albert Einstein*, published by Princeton University Press. He is the editor with Jürgen Renn of *Albert Einstein/Mileva Marić: The Love Letters*.

Philip Glass is a composer and performer who has created numerous works for the theatre, film and dance, and 21 operas including *Einstein on the Beach*.

Max Jammer is professor of physics emeritus and former rector of Bar-Ilan University in Israel. His books include *Concepts of Space*, with a preface by Albert Einstein, whom he knew personally, *The Philosophy of Quantum Mechanics* and *Einstein and Religion*.

Sir Joseph Rotblat is the only physicist to have resigned on ethical grounds from the Manhattan Project to make the atomic bomb; he later helped to found the Pugwash movement to control nuclear weapons. In 1992 he was awarded the Albert Einstein Peace Prize, and in 1995 he and Pugwash were jointly awarded the Nobel prize for peace. Born in Poland, he left in 1939 and became a British citizen in 1946. He was knighted in 1998.

I. Bernard Cohen was the Victor S. Thomas professor emeritus of the history of science at Harvard University until his death in 2003. The last person to interview Albert Einstein, he pioneered the study of the history of science in the United States, and in 1999 he produced the first English translation of Sir Isaac Newton's *Principia Mathematica* since 1729.

Sir Arthur C. Clarke is a science-fiction writer who has written over 70 books of fiction and non-fiction and co-wrote with Stanley Kubrick the screenplay of the film *2001: A Space Odyssey*. A former scientific researcher, in 1945 he conceived the idea of the communications satellite. He was knighted in 1998.

INDEX

Numbers in *italic* refer to illustrations. Einstein's contributions to physics are indexed under the relevant scientific term, e.g. relativity, or laser. His relationships with individuals are indexed under the individual's name, e.g. Bohr, Niels, or Einstein, Elsa.

CREDITS

"Preface" by Freeman Dyson: © Freeman Dyson 2005.

"Autobiographical Notes" by Albert Einstein: Reprinted by permission of Open Court Publishing Company, a division of Carus Publishing Company, Peru, IL, from *Albert Einstein: Philosopher-Scientist* edited by Paul A. Schilpp, copyright © 1988.

"A Brief History of Relativity" by Stephen Hawking: Reprinted by arrangement with Writers House, LLC, as agent for Stephen Hawking. Copyright © 2001 by Stephen Hawking.

"Varying *c*: Vodka without Alcohol?" by João Magueijo: © João Magueijo 2005.

"Einstein's Search for Unification" by Steven Weinberg: © Steven Weinberg 2005.

"Einstein's Scientific Legacy" by Philip Anderson: © Philip Anderson 2005.

"Einstein's Love Letters" by Robert Schulmann: © Robert Schulmann 2005.

"Einstein and Music" by Philip Glass: © Philip Glass 2005.

"Einstein on Religion, Judaism and Zionism" by Max Jammer: © Max Jammer 2005.

"Einstein's Quest for Global Peace" by Joseph Rotblat: © Joseph Rotblat 2005.

"Einstein's Last Interview" by I. Bernard Cohen: Reprinted with permission. Copyright © 1955 by Scientific American, Inc. All rights reserved.

"Einstein: Twentieth-Century Icon" by Arthur C. Clarke: © 2005 Arthur C. Clarke.

"Einstein's Last Interview" by I. Bernard Cohen was first published as "An Interview with Einstein," *Scientific American*, July 1955: 68-73.

"God and Einstein" was included in Arthur C. Clarke's collection of essays, *Report on Planet Three* (1972).

All images are courtesy of The Albert Einstein Archives, The Hebrew University of Jerusalem ('AEA' for short) except as follows:
[t=top, m=middle, b=bottom]
[AIPESVA=AIP Emilio Segrè Visual Archives]

Endpapers, 12: Courtesy of the Archives, California Institute of Technology
2–3: Princeton University Library.
5, 8, 11, 152, 161, 179, 238: © Bettmann/CORBIS.
7, 23, 58, 116, 139, 143, 168, 204, 214, 217, 218, 225: Getty Images.
10: Cartoon by Low/courtesy Caltech Archives.
14b, 162: akg-images.
17, 25, 106, 193: Science Photo Library.
18, 39b, 47, 91, 166, 187, 198–99: © The Hebrew University of Jerusalem.
19, 43, 117t, 219: akg-images/Erich Lessing.
20: Omikron/Science Photo Library.
22: Cordelia Molloy/Science Photo Library.
24, 44b, 56, 75m, 77b, 113, 119, 223: David Woodroffe.
36: Stadtarchiv Ulm.
37: Image Archive ETH-Bibliothek, Zurich.
42: Photograph by Lucien Chavan, courtesy AEA.
44t: Photography by Elmer Taylor, courtesy AIPESVA.
44m: AIPESVA/Original from Case Western Reserve University.
52: © CORBIS.
53, 135: © Underwood & Underwood/CORBIS.
54–55, 67: © The Jewish National & University Library, Jerusalem. Schwadron Collection.
57: American Institute of Physics/Science Photo Library.
59, 76: Science Museum.
60, 72, 137, 172: Leo Baeck Institute, New York.
63t: Henri Manuel, courtesy AIPESVA Physics Today Collection.
63b: W.F. Meggers Collection/American Institute of Physics/Science Photo Library.
64: Ullstein Bilderdienst, courtesy AIPESVA.
65: Deutsches Museum.
70: H.S. Lorentz, A. Einstein, H. Minkowski, *Das Relativitätsprinzip*, 1915, courtesy AIPESVA.
71, 94b: Science Museum Pictorial.
74: William G. Hartenstein.
75b: Don Harley.
79: Photograph by J.F. Langhans, courtesy AEA.
81, 165, 184: ullstein—ullstein bild.
88: Photograph by Paul Ehrenfest, courtesy AIPESVA.
89, 133b: Photograph by Queen Elisabeth (Belgium)/Archive du Palais Royal, courtesy AEA.
90: Prof. Peter Fowler / Science Photo Library.
92: Indian National Council of Science Museums, courtesy AIPESVA.
93, 192: AIPESVA, Brittle Books Collection.
94t: Max-Planck-Institut für Physik, courtesy AIPESVA. (Gift of Max-Planck-Institute via David Cassidy).
96: Max-Planck-Institute für Physik, courtesy AIPESVA, Born Collection.
100: Photograph by Paul Ehrenfest, courtesy AIPESVA.
101: Photograph by Ferdinand Schmutzer, courtesy AEA.
104, 129: © Lucien Aigner/CORBIS.
108: AIPESVA. Archive for the History of Quantum Physics.
109, 112: CERN/Science Photo Library.
110: David Nunuk/Science Photo Library.
111: Space Telescope Science Institute/NASA/Science Photo Library.
114: Photo by Associated Press/courtesy Caltech Archives.
117b: Meggers Gallery/American Institute of Physics/Science Photo Library.
121, 183, 208: Photograph by Alan Windsor Richards, courtesy Bernice Sheasley/AEA.
123: Photograph by Trudi Dallos, courtesy AEA.
124: Photograph by Arthur Johnson, courtesy AEA.
144, 176: Photograph by Martin Harris, courtesy AEA.
153, 230: © CORBIS SYGMA.
160: akg-images/Paul Almasy.
169: Photograph by Martin D'Arcy, courtesy AEA.
174: Third Naval District Public Relations, NY, courtesy AEA.
177, 189: Central Zionist Archives, Jerusalem.
178: Photo donated by Efrat Karmon, courtesy AEA.
185: Courtesy Max Jammer.
191, 194, 212, 220: Time Life Pictures/Getty Images.
195: Copyright © Pittsburgh Post-Gazette, 2005, all rights reserved. Reprinted with Permission.
197, 201: Photograph by Hermann Landshoff, courtesy AEA.
206: Bulletin of the Atomic Scientists, courtesy AIPESVA.
213: Retna Pictures Ltd/Royal Photographic Society/Jarche.
227: Patrick Burns/The New York Times Photo Archives.
234: TIME Magazine ©2005 Time Inc. Reprinted by permission.
239: Arthur Sasse/AFP/Getty Images.

The following sources provided inspiration for the diagrams:
Fig. 2: Hey and Walters (*Einstein's Mirror*).
Fig. 3: Rigden (*Einstein 1905: The Standard of Greatness*).
Fig. 4: Einstein and Infeld (*The Evolution of Physics*).
Fig. 6: *New Scientist* (11 Jan. 2003).
Fig. 7: Hey and Walters (*The New Quantum Universe*).
Fig. 8: Article by Rogers in A. P. French, ed. (*Einstein: A Centenary Volume*).